D1424966

KNOW & GROW
VEGETABLES 2

KNOW & GROW
VEGETABLES 2

J.K.A. Bleasdale, P.J. Salter,
and others

Oxford New York Toronto Melbourne
OXFORD UNIVERSITY PRESS
1982

Oxford University Press, Walton Street, Oxford OX2 6DP
London Glasgow New York Toronto
Delhi Bombay Calcutta Madras Karachi
Kuala Lumpur Singapore Hong Kong Tokyo
Nairobi Dar es Salaam Cape Town
Melbourne Auckland
and associates in
Beirut Berlin Ibadan Mexico City Nicosia

First published as an Oxford University Press paperback 1982,
and simultaneously in a hardback edition

British Library Cataloguing in Publication Data
Salter, P.J.
Know and grow vegetables.
Vol. 2
I. Vegetable gardening
I. Title II. Bleasdale, J.K.A.
635 SB322
ISBN 0-19-217727-3
ISBN 0-19-286017-8 Pbk

Library of Congress Cataloguing in Publication Data
Know and grow vegetables.
(Oxford paperbacks)
Includes index.
1. Vegetable gardening. 2. Vegetables.
I. Bleasdale, J.K.A. II. Salter, P.J. (Peter John) III. Title: Know and grow vegetables two.
SB322.K582 635 81-16977
ISBN 0-19-217727-3 AACR2
ISBN 0-19-286017-8 (pbk.)

Printed in Great Britain by
Richard Clay (The Chaucer Press) Ltd
Bungay, Suffolk

The Contributors

The Editors, who are also contributors, are:

Professor J.K.A.Bleasdale, *Director*,
National Vegetable Research Station,
Wellesbourne, Warwickshire

Dr.P.J.Salter, *Assistant Director*
and Head of Plant Physiology Section,
National Vegetable Research Station

The other contributors are also members of the staff of the National Vegetable Research Station. They are:

Professor N.L.Innes, *Deputy Director*
and Head of Plant Breeding Section

H.A.Roberts
Head of Weeds Section

D.A.Stone
Soil Science Section

Dr.W.G.Tucker
Plant Physiology Section

Contents

1 Choosing a variety

How do *you* choose vegetable varieties for growing in your garden? Do you choose the variety you have grown for the past few years with satisfactory results, or pick one almost at random from the range offered in a catalogue or on a seedsman's display stand? Or do you wisely seek as much information and advice on the subject as possible, including asking neighbouring gardeners who are likely to have similar growing conditions?

Choosing a variety best suited to your particular needs can be difficult, as there is a wealth of varieties for each vegetable crop and there are about thirty different vegetables! Seedsmen generally assess their varieties in special field trials where each variety is subjected to almost identical growing conditions. In this way they are able to build up a picture of the performance and characteristics of each variety. This enables them to decide when to eliminate varieties with faults and also to give a brief description of the main characteristics of the best varieties in their seed catalogues and on seed packets. Generally speaking, however, the seedsmen do not attempt to make detailed comparisons of varieties in their catalogues, but draw special attention to their varieties with new or 'novel' features.

It is impossible to recommend the current varieties that are 'best' for every situation, for these may well differ for different soils, different regions and for different purposes. We shall, however, discuss the characteristics of groups of varieties of most outdoor vegetable crops, and suggest a number that you should try in your garden in comparison with your old favourites. Before doing so, however, we shall indicate sources of information to help you make your choice and describe legislation and official testing that is necessary before a new variety reaches the seedsmen's catalogues.

VEGETABLE TRIALS

Many seed catalogues mention awards made by the Royal Horticultural Society (RHS) who have conducted vegetable trials since 1818. The Society invites breeders and seedsmen from all parts of the world to submit new or improved varieties for testing in their trial grounds, now at Wisley Gardens in Surrey. Each year seven or eight types of vegetables are selected; the frequency of trials for any particular crop depends on the popularity of that crop and the rate of introduction of new varieties.

Wisley is a favourable location but does experience severe frosts. The trials are simple, unreplicated comparisons, and a high standard of cultivation and horticultural skill ensures good plant stands, ample nutrition and moisture on all plots. In each trial at least one currently-recognized variety is included for comparison. Assessments are made by a panel of experts who make awards based on uniformity, growth habit, yield and quality, disease resistance and colour. Only outstanding varieties are awarded a First Class Certificate, whilst Awards of Merit and Highly Commended are made to good varieties in descending order of performance.

These trials have proved to be very valuable to gardeners and growers in indicating the nature and merits of the varieties tested. However, as the number of trials are limited, each crop is tested only every few years, so that it is possible to have excellent new varieties on sale that have not as yet been assessed by the RHS.

Although the RHS is the only organization to undertake variety 'trials' for gardeners, there are official Government-sponsored trials by the National Institute of Agricultural Botany (NIAB), Cambridge, of a number of vegetable crops to provide guidance for commercial growers. Based on the results of these trials NIAB regularly publish Vegetable Growers' Leaflets on the major commercial vegetables, Brussels sprouts, cabbages, maincrop carrots, cauliflowers, butterhead lettuce and bulb onions (spring- and autumn-sown). These leaflets contain a great deal of information about the more important characteristics of varieties that have been included in NIAB vegetable variety performance trials. Such trials are done at a number of different sites each year, often at Experimental Horticulture Stations (EHSs) or Farms (EHFs) of the Agricultural Development Advisory Service (ADAS) of the Ministry of Agriculture, Fisheries and Food (MAFF). The value of the NIAB trials lies in their being done at several sites over several years, and in each trial, each variety is tested in several randomized plots so that a statistical test may be used to compare the different varieties in the trial and give a reliable result. For the so-called minor vegetable crops there are relatively few trials organized by NIAB. Fortunately, MAFF have done trials on crops such as rhubarb, runner beans, sweet corn and marrows at EHSs so that information in their annual reports is often useful in identifying good new varieties.

A leaflet by NIAB entitled *Vegetable Varieties for the Gardener* makes use of the results of their official trials to suggest varieties for the crops that they have tested.

HOW DO NEW VARIETIES GET ONTO THE MARKET?

Nowadays a breeder or seedsman looks to commercial vegetable growers to take up a new variety first, in the knowledge that the sale of seed in small packets to the general public is likely to follow. However, before seed of new varieties can be sold to anyone it must undergo certain official tests, so that buyers are protected from purchasing the same variety under different names. Varieties which pass the criteria of the 'Distinctness, Uniformity and Stability' (DUS) tests are accepted for entry on the National List of Vegetables. They are then a registered variety and can be marketed. DUS tests are conducted by NIAB for the MAFF or by the Agricultural Scientific Services of the Department of Agriculture and Fisheries for Scotland (DAFS). These tests assess the uniqueness of a new variety by ensuring that it is different from other varieties, and that it is sufficiently uniform within the crop and stable from year to year. When a variety has been accepted for inclusion in the National List it is also acceptable for the EEC Common Catalogue so that it can be lawfully marketed in EEC Member Countries. A variety may be marketed in the UK, therefore, even if it is not on the UK National List, provided it is included in the Common Catalogue and has therefore been tested in one of the EEC Member Countries.

It is illegal to sell seed of varieties under the wrong name or which are not on the National List or Common Catalogue. This regulation has undoubtedly done much to rationalize the great number of different names attributed to the same varieties. However, it has to be recognized that varieties will only remain on the official lists for as long as a seed firm is prepared to maintain them. Some old varieties are lost because there is insufficient demand for them to be economically worthwhile for a seedsman to multiply seed stocks. Fortunately steps have been taken at the National Vegetable Research Station (NVRS) to safeguard the loss of valuable breeding material by collecting and storing old, displaced varieties of vegetables in a vegetable Gene Bank.

In addition to National Lists, there is an increasing trend to encourage investment in the breeding of improved varieties by plant variety rights schemes, which are equivalent to 'plant patents'. Varieties entered for Plant Variety Rights are subjected to the same sort of DUS tests as apply to National Lists. Where a variety is accorded rights, the holder of the rights is entitled to a royalty payment on the sale of all seed of that variety for the period that his rights remain in being, subject to a maximum period of 15 years for vegetables and 20 years for potatoes.

VARIETIES BRED AT GOVERNMENT-SPONSORED INSTITUTES

We are often asked, particularly at NVRS Open Days, to provide lists of suppliers of seed of varieties bred at the NVRS and other Government-sponsored research institutes, for example, our butterhead lettuce *Avondefiance* or the cabbage *Celtic* from the Scottish Horticultural Research Institute. It is appropriate, therefore, to explain how such varieties usually reach the gardener as the research institute which breeds them does not market the seed.

At the same time as a vegetable variety undergoes official DUS tests and statutory performance trials, the breeder (whether private or public) is usually concerned with building up seed stocks. The breeder produces small quantities of élite seed, officially known as 'pre-basic'. In the case of varieties bred at Government-sponsored institutes this seed is passed to the National Seed Development Organisation (NSDO) at Cambridge, who produce from it larger quantities of 'basic seed'. Great care is taken in the production of both pre-basic and basic seed, and multiplications are frequently made in glasshouses or polythene tunnels where strict control over pollination is possible.

NSDO is responsible for the marketing of the variety and sell basic seed to commercial seed houses, who multiply it further, often in countries where a hot, dry climate is more suitable for seed production. The seed companies use this final generation of seed for sales to the general public. There are variations of this scheme but in general the same overall pattern is followed.

Profits from NSDO's income, which is derived from both sales of basic seed and from royalties on sales made by seed companies, are returned to the Exchequer as a return on the State's investment in plant breeding work in the public sector. Seed of Government-bred varieties is available from most, though not all, seed companies.

BREEDING FOR IMPROVED QUALITY

A commonly-held opinion, particularly among older gardeners, is that new varieties of vegetables have been bred only to increase yield and this has been achieved at the expense of quality, especially flavour. Quality characteristics such as colour, texture and flavour have received considerable attention by vegetable breeders, and in recent years, beetroot, carrots and sweet corn have all been improved for sweetness. New celery varieties are less pithy, green French bean varieties are now stringless and improvements in this direction are also being made in runner beans. Many new varieties of carrots, beetroot, tomatoes, green beans and peas have improved colour and appearance, and the breeding of bolting-resistant varieties of celery, carrots, beetroot and cabbage has done much to improve the quality of these crops.

Flavour is extremely difficult to define and characterize, and tastes vary. As growing conditions affect the flavour of a variety, the breeder has not found it easy to quantify flavour differences, though more and more breeders are turning their attention to flavour components in vegetables and major successes have been achieved, such as the elimination of the bitterness in cucumbers. Until techniques are developed to quantify flavour components the plant breeder must still rely on subjective assessments, usually by taste panels. We remain unconvinced that there has been as serious a loss in flavour in new varieties as is sometimes claimed, and would argue that only the individual, by trying a number of old and new varieties grown side by side, under the same conditions, is best placed to judge which flavour pleases him or her most.

SELF-POLLINATED VARIETIES

When a plant sets seed with its own pollen this is called self-fertilization, and where this is the normal method of reproduction the plant is called an inbreeder. In inbreeding crops self-fertilization usually takes place without the assistance of insect-pollinators, although there are exceptions. For example, visits by bumble- or honey-bees to the flowers of runner beans are necessary to get pods to set. Runner beans are perfectly capable of setting pods containing seeds by self-fertilization but their flower structure is such that it needs a bee to 'trip' the flower so that pollen grains land on the surface of the stigma and penetrate to the ovules. Dwarf French beans have no such problems and set their own seed quite readily without pollination by insects, because pollen is deposited directly onto the stigma surface within the flower.

Among inbreeders are lettuce, peas and tomatoes. Lettuce is so strongly self-fertile that breeders have difficulty in making their crosses, while French beans, peas and tomatoes can cross-fertilize if cross pollen is present, usually having been carried there by bees, when the female parts of the flowers are receptive. Because of this ability to set cross-seed, seedsmen usually grow their seed of these crops in isolation.

When breeding new self-pollinated varieties the breeder usually starts with a variable population, often created by crossing together different varieties. After repeatedly selecting large numbers of single plants and testing their offspring for seven or eight generations, seed of each of the most promising offspring is bulked for testing as potential new varieties.

CROSS-POLLINATED VARIETIES

A number of vegetable crops, particularly brassicas, onions and carrots, are either cross-pollinating or set much more seed of better quality when cross-pollinated. Such crops can, and do set varying quantities of self-fertilized seed but usually these seeds produce plants that are weaker than those from crossed-seed. Cross-pollinating plants have special biological mechanisms in their flower parts that encourage cross-mating and usually insect pollinators are involved in moving pollen between plants. As usual, there are exceptions and both sweet corn and beetroot are wind-pollinated. Until recently, varieties of cross-pollinated crops were maintained by growing a large population of each variety in isolation so that crossing took place between plants within the variety but there was no contamination from other varieties. The advantage of crossed, or hybrid, seed in cross-pollinating crops has been exploited by plant breeders to breed hybrid varieties which are superior to open-pollinated varieties.

HYBRID VARIETIES

Hybrid breeding is highly specialized and much more costly than older, more traditional methods (see Fig. 1.1). In the first place, the breeder has to devise and use special techniques to force a plant that is usually cross-fertilized to set self-fertilized seed. Inevitably the amount of seed that is set is much less than that obtained from cross-fertilization, and the offspring from the self-fertilized plant is much less vigorous, or suffers from 'inbreeding depression'. Forced inbreeding over several generations, accompanied by rigorous selection and elimination of any plants that show defects, provides extremely uniform, though somewhat weak inbreds. Intercrossing of pairs of uniform inbreds gives hybrids which are of better quality and higher yielding than the inbred parents from which they were derived, and are equally uniform. Unfortunately not all pairs of inbreds combine well and it is usually necessary to judge the performance of hybrids by trial crosses, thus increasing the cost of the breeding programme. As inbred seed is costly to produce, and yields of hybrid seed (which sets on, and is harvested from weak inbred parents) are relatively low, the cost of hybrid seed is high. Breeders have tried various methods to increase hybrid seed production and more use is being made of double-cross hybrids, where hybrid from (A × B) is crossed with hybrid (C × D). Unfortunately these double-cross hybrids are rarely as uniform as single crosses but are often still superior to the open-pollinated varieties from which they were derived.

Over the last ten years hybrid varieties of many crops have

become available. Hybrid seed is always the most expensive but is it necessarily the best? The value of hybrids depends on the crop. With Brussels sprouts and cabbages it is generally true to say that hybrids are much better than nearly all of the old varieties which they have replaced – certainly F_1 hybrid sprouts have done much to increase marketable yields for commercial growers, and to make sprout buttons more uniform for the processor. However, it is important to realize that not all hybrids are the answer to a gardener's prayer, even in cross-pollinating crops, for some non-hybrids are well worth growing. For example, two excellent cabbages, the spring cabbage *Avon Crest*, which is very resistant to bolting, and the extremely uniform January King type *Avon Coronet*, bred at the NVRS, are non-hybrids.

Among carrots and bulb onions hybrid varieties are beginning to make their appearance but it will be necessary to await the results of performance trials to estimate their superiority – if any – over non-hybrids. Because inbreeding depression is extremely severe in onions and carrots, the parents used to make hybrids have only been inbred for two or three generations and are not completely uniform so that their hybrids are still somewhat variable. In the self-pollinating crops, such as lettuce, French beans and peas, hybrids are extremely difficult to breed and usually their advantage over self-pollinated varieties is so small that it is not economically worthwhile to embark on hybrid breeding programmes. However, there is always the exception and although tomatoes readily self-pollinate, F_1 hybrid tomatoes do generally outyield self-bred varieties and often have resistance to several diseases. As it is relatively simple to produce large numbers of hybrid seed by hand-pollination of tomatoes F_1 hybrids have understandably become popular, and among outdoor bush tomatoes, *Sleaford Abundance* and *Alfresco* are both hybrids that give excellent yields. With a renewal of commercial interest in outdoor tomatoes some of the non-F_1 hybrid bush selections from Canadian and Russian material seem likely to do well.

HOW IS SEED PRODUCED?

Most of the vegetable seed sold in the UK is produced overseas, although some seed production, particularly of brassicas, is still done in Essex. A seedsman usually hedges his bets by placing contracts with two or more seed producers, as inevitably there are crop losses from pests and disease attacks, flooding, and invasion by unwanted livestock or rabbits and hares. With partial seed crop failure, demand can outstrip supply, particularly if a variety has a major outlet among commercial growers. It is not surprising therefore to find occasionally that a seedsman is 'out-of-stock'.

As a generalization, all seed production is best left to the specialist, especially as seed-borne diseases such as halo-blight in runner beans, mosaic virus in lettuce, *Alternaria* disease in brassicas, and neck-rot in onions, can cause serious problems to the amateur. In addition, with cross-pollinated crops, and even those that are largely self-pollinating, there may be contamination because of inadequate isolation of seeding crops, so that unwanted crosses from other varieties or even from wild species occur. Apart from disease problems, the seed obtained in our extremely variable climate may be of poor quality and low germination.

Among the cross-pollinated crops the low frequency of pollinators such as honey-bees and low temperatures at flowering can lead to problems with seed-set. With biennial crops there is the added difficulty of crops occupying land for two years, thus increasing the possibility of the build-up of pests and diseases. With such biennials there is inevitably a tendency for the amateur to harvest seed from plants that most readily 'run to seed', or 'bolt' (see page 179). When this happens you can unwittingly make your next crop much more susceptible to bolting.

Temptation to produce and keep seed from a hybrid variety should be avoided, as plants raised from such seed will not provide replicas of the hybrid parent from which they derive, but will be a highly variable population containing both good and bad types known to the scientist as segregants.

In view of the difficulties in multiplying your own seed, and with the possible exception of the 'easier' crops such as peas and beans (and even these have their problems), seed production is best left to the specialist.

To minimize expenditure on seed it is best to store any you don't use employing the method described in Chapter 2 of the previous book *Know and Grow Vegetables*.

VARIETIES

In their leaflets, NIAB specify the average number of days to maturity of different varieties of vegetables. These figures can vary from year to year and place to place and should be taken as no more than a guide; the important aspect of the information is the relative order in which varieties mature as this does not usually vary. All such information has been taken into account here. Where continuity of production is sought (see Chapter 2) an attempt is made to include at least two varieties for a particular harvest period. In the varieties listed here emphasis is given to 'all-purpose' rather than the specialized varieties often sought by the commercial grower. Sometimes the commercial grower will see defects in a variety that are unlikely to affect a variety's attractiveness to a gardener. For example, for several years the butterhead lettuce variety *Avondefiance* has been the standard against which new varieties have been judged in NIAB's trials. This variety, which is resistant to lettuce root aphid, to tipburn disease and to several races of downy mildew, gives excellent crops from June to October. Unfortunately, from a commercial viewpoint, it has a somewhat pointed base, which makes it difficult to pack into boxes for market, and if exposed upside down on a supermarket shelf, its 'butt' turns brown. Both these 'defects', which are unlikely to be of any importance to the gardener, are leading to it being replaced with other varieties by the commercial grower.

Occasionally a variety gives outstanding yields and quality under very fertile conditions but suffers more adversely, relative to other varieties, under poor growing conditions. Some hybrid varieties of Brussels sprouts certainly behave in this way; such varieties have not been included in our lists. It is recognized that some companies and individuals specialize in seed to produce mammoth plants of, for example, onions, for horticultural shows. No attempt has been made to include such strains or varieties, although the listed varieties should give a high quality product.

The cultural notes that are usually to be found on a packet of seed of a particular variety should be taken as guidelines to ensure that a variety is not sown at the wrong time or for the wrong purpose. For example, there is no point in attempting to grow the glasshouse lettuce variety *Dandie* as a summer garden lettuce or to attempt to grow Roscoff broccolis (winter cauliflower) in the north of England, as they only do well in the milder parts of southern Britain. Varieties will only be at their best if sown at the correct time and at the optimum spacing, and fertile conditions are required to ensure best results. Gardeners raising plants for agricultural shows may deviate from the sowing dates indicated in this chapter by raising seedlings earlier than specified under heated glass or indoors. In listing a number of varieties in the following pages, we have taken into account the likely availability of seed to gardeners, as quite often seed of a new variety may be available in bulk to commercial growers but not in small packets.

As there is little information from comparative trials on asparagus, kohl-rabi and shallots, no attempt is made to list varieties for these crops. Asparagus crowns are often available from local growers, who have selected and maintained their own strains. Over the next few years new French hybrids are likely to become available.

Kohl-rabi, a brassica grown for its swollen, bulbous stem which develops just above ground level, has a flavour between cabbage and turnip. It should be harvested when young and tender. Both purple- and white-skinned varieties have white flesh.

Shallots are a hardy, perennial onion whose bulb splits and divides to form six to twenty new shallots which are excellent for pickling. Particularly popular in more northern regions,

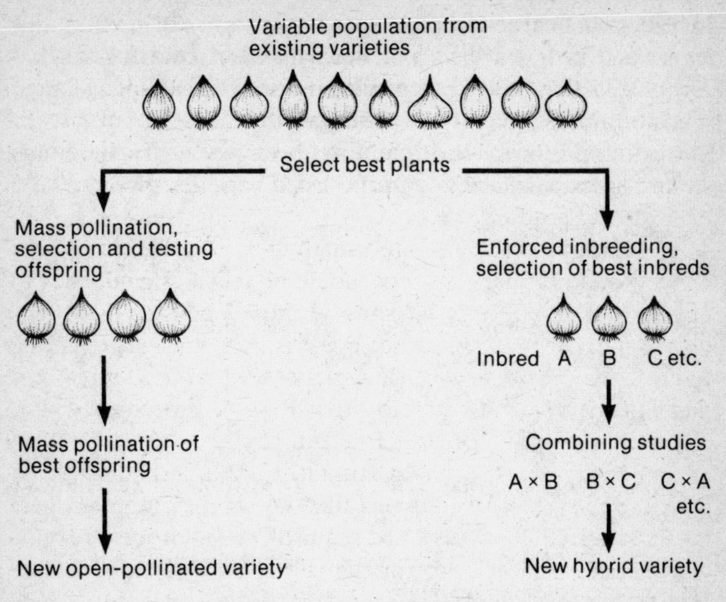

Fig. 1.1. Simplified version of breeding open-pollinated and hybrid varieties in cross-pollinated vegetables.

red and yellow varieties are available as bulbs, often as virus-free stocks, from a number of suppliers. Many gardeners maintain their own stocks but it is advisable to replace these from time to time if the proportion of stunted or diseased plants builds up.

Where a variety has an asterisk it is, as yet, marketed by one or two seedsmen only but is likely to become more widely available over the next few years.

Where necessary, sowing times have been given for particular varieties; for a full description of sowing and harvest dates, see Chapter 2.

The different shapes that are obtainable in carrots, onions, parsnips, and red beet are illustrated in Fig. 1.2.

Beetroot

The breeding of bolting-resistant, globe-shaped varieties has made possible earlier sowings to produce deep-red, small, tender roots which are excellent hot as well as cold for salads. *Detroit* selections are available for main cropping but the early, bolting-resistant varieties are also suitable. The first two varieties listed have done well in NIAB trials, while the other one has given satisfactory results over a number of years.

Avonearly, Boltardy Both are early, globe-shaped (Detroit-type) varieties with resistance to bolting and red flesh that is almost free from white rings.

Cheltenham Green Top A broad-shouldered, long-rooted, medium-size variety of good quality, usually grown for winter storage.

Borecole or Curly kale

In this crop, which is popular among more northerly gardeners perhaps because it is one of the most frost-resistant vegetables, the following varieties have done well in recent years.

Fribor* An F_1 hybrid of medium height with fine, curly, deep-green leaves for harvesting November to February.

Pentland Brig A vigorous kale producing an abundance of young leafy shoots in early spring; more frost-hardy than purple-sprouting broccoli.

Broad beans

There are three main types of broad beans, Seville, Longpods, and Windsors, which can be further subdivided on the basis of white or green seed colour. The Seville types are hardier and therefore more suitable for autumn or winter sowing. The Longpods have slender pods with up to eight kidney-shaped seeds, while Windsors have shorter pods with four or five seeds, which are flat and circular. There is little difference between white and green seeds in their flavour, though some think that green seeds are more tender.

A fourth type, the Dwarf or Fan-podded, are bushier, dwarf plants which mature quickly, carrying many small pods, each with three seeds, and varieties of this type are particularly useful for the small garden.

The varieties that are listed according to their sowing period have given good results over many years.

Seville

Aquadulce Claudia A tall white-seeded variety for sowing in late October/November; has a good flavour.

Longpods

Imperial Green Longpod A tall, very long-podded, green-seeded variety for February to April sowing; good for freezing.

Exhibition Longpod The traditional synonym of the variety *Conqueror*, it is a heavy cropping white-seeded bean with good flavour for spring sowing.

Windsors

White Windsor, Green Windsor Seedsmen's selections from these two varieties give late yields from sowings during March-May.

Dwarf (*Fan-podded*)

The Sutton A white-seeded variety for sowing under cloches in November/December or in the open from February to July.

French beans

The listed varieties have done well in trials done by the NIAB, the Processors and Growers Research Organization, Peterborough, and the Campden Food Preservation Research Organization.

Bina* A variety with flat, fleshy pods like a runner bean but free from string. Very good for slicing.

Glamis* A stringless, early-maturing variety of good flavour bred especially for Scottish conditions. Has brown seed and is likely to be replaced by its white-seeded successor *Glenlyon.**

Kinghorn Wax* A waxy-podded type with fleshy, stringless pods.

Loch Ness A stringless variety of upright growth with long, round, straight pods which are suitable for slicing and chopping.

Pros A stringless variety with slightly curved, medium-green pods that are oval in cross section. Very good for cooking or freezing as whole pods.

Tendergreen A popular heavy cropper with fleshy, stringless, round pods which freeze well.

Runner beans

Most gardeners grow runner beans as a climber with supports, thus ensuring a heavier yield of cleaner pods than is obtained from 'pinching-out' the crop by removal of the growing point. The dwarf variety *Hammonds-Dwarf Scarlet* is available but has the disadvantage that many of its pods are damaged through contact with the soil; 'pinched-out' *Kelvedon Marvel* has given good results as a dwarf crop.

Some white-flowered varieties are simply white-flowered variants of scarlet-flowered varieties (for example *White Achievement*), but others (for example *Fry**) are quite distinct from scarlet forms. Several relatively stringless varieties have been introduced recently, including *Desiree**, *Red Knight** and *Mergoles**. Among the varieties listed here are those that have done well in trials at ADAS EHSs, or in trials at the NVRS.

Achievement, Enorma, Prizewinner, Scarlet Emperor, Streamline All good maincrop varieties with long, straight, green pods.

Fry* A white-seeded variety with white flowers which yields well late in the season; pods relatively stringless.

Kelvedon Marvel Sometimes known as *Kelvedon Wonder*, this is an early variety with shortish pods. Suitable for 'pinching-out' under cloches and also as a 'pinched' outdoor crop.

Sunset* A variety with pale-pink flowers that produces a good crop of medium-length pods.

Broccoli, sprouting

There are as yet no named varieties of this excellent garden crop, which provides a welcome fresh vegetable during March/May. Seed is usually sold as either purple- or white-sprouting broccoli, both of which are subdivided into early- or late-. There is not a marked difference between early- and late-purple-sprouting in terms of maturity which together provide spears in March and April. However, within white-sprouting the early strains are distinctly different from late, giving spears in March/April while the late crop provides spears or small, cauliflower-like heads in May.

Nine-star perennial broccoli has a central head surrounded by several smaller heads and may be a form of late-white-sprouting broccoli. Although plants may, with care, last several years, their yields are not maintained and there is a danger of their acting as foci for the carry-over of pests and diseases.

A 'broccoli' variety currently sold as a novelty is *Romanesco*, which is really a primitive form of pale-green cauliflower originating in Italy. Somewhat variable, this variety has a texture and taste different to other broccolis and cauliflowers.

Brussels sprouts

As the NIAB do annual trials at several sites on Brussels sprout varieties, and trials are also done at the NVRS, there is a lot of information on which to base decisions about which varieties to grow.

The efforts of plant breeders have produced a crop that is 'in-season' over a period extending from September to March. With the exception of early hybrids, most F_1s now 'hold' their buttons well and produce sprouts of much better quality than open-pollinated varieties, so only hybrids are recommended here. Although a harvest period for maximum yield is given, most varieties will continue to produce some harvestable sprouts beyond that period. The varieties listed according to their maturity, or harvest period, all give high yields.

Early (September/November)

Peer Gynt Short to medium plants with medium-sized, smooth, solid sprouts.

Topscore Short to medium plants producing good quality oval to round, medium-sized sprouts.

Cor Formerly known as *Valiant*, this heavy yielding variety has medium, slightly elongated solid sprouts which hold well. Tends to overlap with mid-season sprouts.

Mid-season (*November/January*)

Citadel A medium-tall variety with good quality round sprouts.

Perfect Line A consistent performer with heavy yields of medium to large, rather oval, dark green sprouts which hold well.

Widgeon* A British-bred variety with heavy yields of excellent quality.

Ormavon* A British-bred variety with medium to large sprouts of good quality; a cabbage-like head on the plant is also consumable.

Welland* A British-bred variety with larger sprouts and good resistance to powdery mildew.

Late (*December/March*)

Achilles A British-bred variety with small- to medium-sized, mid- to dark-green sprouts of excellent quality which hold well. A tendency to fall over in some localities.

Rampart* A tall, high-yielding variety producing medium to large sprouts of good quality. Although cropping starts in the mid-season period, it holds its buttons well.

Sigmund* A hybrid with medium to tall plants producing small- to medium-sized sprouts of good quality.

Cabbages

By careful planning in the medium-sized to large garden or allotment, it should be possible to meet a household's needs all the year round by growing different types of cabbage. Varieties are listed according to maturity, or harvest period, and most have done well in NIAB trials.

Spring (*March/May*)

Avon Crest An early-maturing pointed-head cabbage of good uniformity that can be used as spring greens or as a late-maturing headed cabbage with medium to large, firm hearts. It is extremely resistant to bolting and can also be sown in early July to produce an autumn cabbage.

Durham Early Another dark-green variety of good uniformity; it is slightly susceptible to bolting and is best used as greens. It produces conical, medium-sized heads.

First Early Market 218 A dark-green variety of moderate to good uniformity that is very suitable for greens and can also be allowed to produce large, pointed heads.

Harbinger An early-maturing variety with small, pale- to medium-green hearts. It is somewhat susceptible to bolting and can suffer from frost damage.

Myatts Offenham Compacta* A medium- to dark-green variety for April/May maturity, it has medium to large heads and is of good uniformity and bolting resistance.

Summer (June/July)

Derby Day* A round cabbage which can be brought to maturity early by sowing under glass in February. Of moderate to good uniformity it has small, blue-green heads.

Hispi A popular early-maturing F_1 hybrid with good, uniform, dark-green pointed heads.

Greyhound A popular, quick-maturing variety, with small- to medium-sized, pointed heads.

Marner Allfruh* An early-maturing variety with medium mid-green, round to oval heads of good uniformity. Has a dense head with a low percentage of internal stalk. Also suitable for late sowing to harvest in August.

August/September

Market Topper* An F_1 hybrid of uniform, medium-sized, blue-green heads that will stand without deteriorating for about a month.

Marner Allfruh* By later sowing this variety can be harvested in August (see June/July varieties above).

Minicole A popular hybrid with small, oval, solid heads that stand for up to three months without serious deterioration.

Stonehead An F_1 hybrid producing uniform, solid, round, small heads that will stand for about a month without deteriorating.

Autumn (September/November)

Hisepta* An F_1 hybrid with large, round, grey-green heads that will stand for about two months.

Choosing a variety

Winter (November/February)

White

Jupiter* An F_1 hybrid with round, uniform heads with dark-green outer foliage and good standing ability. Suitable for use as fresh cabbage, for coleslaw, or for storage in frost-free, cool conditions from January to March.

Hidena* Another F_1 hybrid with round to oval heads that have dark, grey-green outer leaves; it has good standing ability. Also suitable for use as fresh cabbage, coleslaw or storage.

January King

Avon Coronet* A very early type with small, uniform heads maturing November/December; good standing ability.

Aquarius* An F_1 hybrid with pale, grey-green, small heads which will stand up to two months without serious deterioration.

December/February

Savoy × White Cabbage Hybrids

Celtic An outstanding F_1 with uniform, round, medium-green heads which have solid, dense, white hearts; it will usually stand up to two months without deteriorating.

Celsa* A later F_1 than *Celtic*, it is also uniform with round, darker-green heads and more blistered leaves. It has less dense heads and will not stand as long after maturity as *Celtic*.

Savoys

Ice Queen An F_1 hybrid with flat, round, medium-green, uniform heads with fine blisters; standing ability up to two months.

Winter King A late-maturing variety with flat, round, blue-green heads that are sometimes small and variable.

Calabrese (or green-sprouting broccoli)

The harvest season for calabrese, often known as poor man's asparagus, is late summer and autumn. When the central head has been harvested, the later development of side-shoots gives extra crop. Although hybrids have almost replaced open-pollinated varieties, for a greater spread of harvest, older varieties such as *Green Sprouting* or *Italian Sprouting* may attract some gardeners.

The varieties listed have done well in trials at the NVRS.

Corvet A medium-early hybrid with uniform, big, round heads, it gives high yields of good quality.

Express Corona A very early hybrid which freely develops side-shoots after the central head has been harvested.

Green Comet An early hybrid giving a high yield of large deep-green heads and relatively few lateral shoots.

Green Duke A dwarf-growing hybrid maturing about the same time as *Green Comet*, it quickly develops side-shoots after harvest of the main head.

Carrots

Modern varieties of carrots have been selected for deeper orange colour, better texture and elimination of quality defects such as green-topped, split and forked roots.

There are a number of different types, the shapes of which are illustrated in Fig. 1.2; those that are available to gardeners are described briefly in the order of their maturity or harvest period. New shapes are now being derived from selections made in crosses of the main types, and further improvements are being made in colour and root quality. Among older varieties of carrots, *St. Valery*, which is also known as *New Red Intermediate*, has a long, tapered root of good colour and texture and is still used by gardeners for exhibition purposes.

Seedsmen have traditionally made their own selections and until recently there was a proliferation of different names, many of which were synonyms for the same variety. The National List and Common Catalogue for carrots now contain far fewer names, but there are still quite a number of approved selections that have been made by different breeders and seedsmen from recognized varieties. Several seedsmen market very good varieties under group names, for example, *Amsterdam Forcing, Early Nantes* and *Chantenay Red-Cored*,

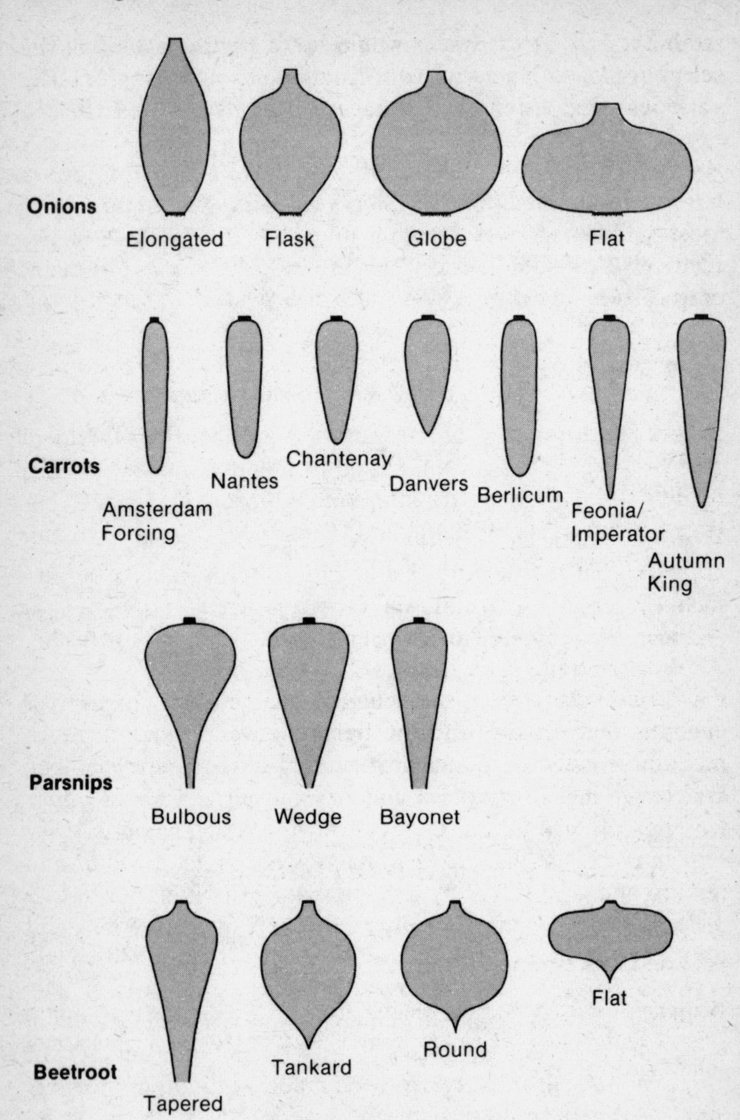

Fig. 1.2. Key to shapes of bulb and root crops.

so that a useful experiment would be to compare some of the selections listed here with your traditional source of seed. The varieties listed are those that have done well in NIAB trials.

Amsterdam Forcing This stump-rooted (rounded tips) group has small- to medium-size, slender, cylindrical roots and is most suitable for early forcing under glass or plastic. It has relatively little foliage and yields are lower than those of main-crop carrots.

Colora* Has a deep, attractive, internal colour with little greening within the roots.

Nantes The roots are of medium-size and have a cylindrical stump-shape, usually with a slight shoulder and sometimes tending to taper. They are intermediate in size and foliage between *Amsterdam Forcing* and *Berlicum*. Like *Amsterdam Forcing* they can be used for forcing.

Nantes – Express* A good yielder with good flesh, but paler core colour and a tendency to internal greening.

Chantenay This group is widely grown by British growers for canning, dehydrating, and the fresh market. The roots are of medium-size and have a conical, stump-shape. The foliage size is between that of *Berlicum* and *Autumn King*.

Chantenay Red-Cored Several seedsmen's selections have done well in NIAB trials, among them *Royal Chantenay*.

Berlicum Large, cylindrical or slightly tapering roots have rounded tips (stump-rooted).

Berlicum – Berjo* A good yielder that has produced roots with a good flesh colour, and a low level of splitting but has a tendency to internal greening.

Autumn King This group has rough-skinned roots which are very large, and have a tapering stump-rooted shape, usually with bold shoulders. Varieties have large, vigorous foliage, mature late and give heavy yields of roots with inferior quality to that of the Chantenays.

Autumn King – Giganta* Gives very high yields of large roots with good flesh colour and low levels of splitting but has a high level of internal greening.

Autumn King – Vita Longa Gives high yields of good core and flesh colour.

Cauliflowers

Cauliflowers were developed in northern Europe to give summer-maturing types, and in maritime regions of north-western Europe including England, to give winter and spring types. The introduction of cauliflowers to Australia in the nineteenth century resulted in further types being developed from which has recently been bred a series of varieties maturing in the autumn.

Among cauliflowers there is a wide range of maturity types and it is possible to produce cauliflowers commercially throughout the year in the UK. However, the curd of the cauliflower is extremely susceptible to frost and winter-maturing forms can only be grown successfully in relatively frost-free areas, such as Cornwall and Pembrokeshire. These winter cauliflowers should not be confused with the winter-hardy or spring-maturing cauliflowers which are ready from March to early May, and can be grown in a much wider range of conditions.

New varieties are fast replacing the older traditional ones such as the Autumn Giant *Veitch's Self-Protecting*. Other older varieties like *All the Year Round*, still have a place in the garden. Although the old English Winter varieties such as *St. George* are of poorer quality than the Angers varieties maturing in late May, they are much hardier.

A number of new varieties have done very well in NIAB trials where yield, curd size, and quality of curd (which includes freedom from defects such as looseness, bracts, riciness and discolouration) have all been taken into account. The best of these varieties are listed according to their harvest period.

Early summer (June/July)

Mechelse – Classic*, Mechelse – Delta Both varieties mature about the end of June.

Dominant Matures in July.

Late summer (July/August)

Nevada This variety can be used for successional sowings to give early summer, late summer and early autumn crops.

Dok Formerly known as *Elgon*, this variety gives best results when grown for maturity in late August and September.

Autumn (August/December)

Flora Blanca A number of seedsmen's selections of this variety (synonym *Autumn Glory*), have given very solid white curds in August/October.

Orco* This variety (formerly *Talbion*) is for cutting in October.

Barrier Reef Matures from October to November.

Winter (December/April)

Often known as Roscoff broccoli these are only suitable for south-west Britain. The St. series covers the complete season for this crop and the following varieties have done best in NIAB trials.

St. Agnes* Matures late December/January.

St. Buryan* Matures February/early March.

St. Keverne* Matures late March/April.

Spring (March/May)

Angers No 2 – Westmarsh Early* Matures in March but is not as winter-hardy as the *Walcheren* varieties listed here.

Walcheren Winter-Armado April* Matures in April.

Walcheren Winter-Markanta Matures in late April.

Walcheren Winter-Birchington* Matures in late April/early May.

Angers No 5 – Summer Snow* Matures in late May, but its quality is inferior to that of the *Walcherens* listed above.

Celery

Self-blanching and green varieties of celery should be used when summer cropping is required, but if sown too early they tend to bolt. They are also rapidly replacing the traditional 'trenching' types for cropping later in the year, but do not survive hard frosts. Thus for winter cropping a 'trench' variety is essential.

Some of the varieties listed here are favoured by commercial growers, while others have given good results over many years.

Traditional or trench

Giant Pink Gives solid, crisp stalks of excellent flavour.

Giant Red A hardy variety with dark-red outer stems with blanched centres.

Giant White Although it produces very high quality stems, this variety needs good conditions and care.

Self-blanching

Avonpearl Of excellent quality and quick growth, it will bolt if sown too early.

Golden Self-Blanching Has compact heads with yellow stems; is ready for use from August onwards.

Lathom Self-Blanching This has become popular in commerce. It is early, crisp, with a good flavour and has very good bolting-resistance.

Green

American Green – Greensnap This has pale-green, crisp stems of good flavour, but tends to bolt.

Chinese cabbage

The most popular forms of this crop are like cos lettuce in shape and have a mild flavour mid-way between cabbage and celery. They can be eaten like a lettuce or cooked as a cabbage. The crop requires a combination of shortening days and fairly high temperatures to avoid bolting and ensure good growth, therefore sowing is best done in July. Like calabrese, it is best sown *in situ*, as it does not transplant too well from the soil but can be raised in pots or peat blocks for transplanting. It is very useful as a catch crop to follow, say, early potatoes. New varieties from Japan are likely to become more readily available as there is increasing commercial interest in this crop.

So far there have been few trials on this new crop but in those that have been done by ADAS and at the NVRS the following varieties have done well.

Nagaoka F_1 Has large, cylindrical, tight heads and is very uniform.

Tip Top F_1 A quick-maturing variety similar in appearance to a large cos lettuce.

Leeks

For many years, particularly in southern England, this excellent winter-hardy vegetable was regarded as merely a flavouring for soups but has now become much more popular in its own right.

The listed varieties have done well in NIAB trials.

Autumn Mammoth – Argenta* Although maturing in late autumn this attractive leek, which has a non-bulbous shank about 5in. long, is suitable for harvesting in winter.

Autumn Mammoth-Herwina* A winter-maturing variety with a dark-green flag and a non-bulbous shank about 5in. long.

Giant Winter-Catalina A hardy, heavy, thick-shafted variety for winter use.

Lettuce

Within this crop there are three main types: butterheads (which have soft delicate leaves), crispheads (which are usually larger with crisp, wrinkled, succulent leaves), and cos (which have longer leaves that are crisp and sweet). Butterheads and crispheads are sometimes referred to collectively as cabbage types. There are also some loose-leaved types from which the leaves can be regularly picked.

Choosing a variety

Varieties suitable for glasshouses have usually been bred for short-day (winter) conditions and are liable to bolt if grown outdoors in summer.

A number of varieties are promoted on the basis of their resistance to downy mildew. There are a great number of different races or strains of the fungus that causes this disease, so that resistance of a particular variety may 'break down' if the fungal strains in a locality are of the type to overcome a variety's resistance.

The listed varieties have done well in NIAB trials and in experiments at the NVRS.

Butterhead

Avondefiance An extremely reliable variety which withstands dry conditions and high temperatures, it produces a heavy yield of good quality heads, medium to dark green in colour. Resistant to root aphid and to some races of downy mildew.

Hilde II Paler than *Avondefiance*, it has a better-shaped heart but is not so good for harvesting after August.

Reskia* Resistant to some races of mildew, this is a variety that may suffer from bleaching during very hot spells in mid-summer, but offers an alternative to *Hilde II* for harvesting after August.

*Sabine** and *Capitan** are extremely promising new varieties of which seed is still in short supply. Both are resistant to some races of downy mildew and, in addition, *Sabine** is resistant to lettuce root aphid and *Capitan** to lettuce mosaic virus.

Crisphead

Most current crisp varieties suffer from some defects. *Webbs, Wonderful* is an old favourite and the following varieties have given good results under most weather conditions.

Avoncrisp A variety resistant to root aphid and to some races of mildew, it can suffer badly from tipburn.

Great Lakes A variety with a good, large head.

Minetto A variety which forms well-folded, compact heads with few outside leaves.

Cos

Lobjoits Green Cos A dark-green variety with medium to large heads.

Little Gem A small, quick-maturing, sweet variety half-way between cabbage lettuce and cos in shape.

Special purpose

Salad Bowl A bolting-resistant, loose-head variety with endive-like leaves that can be picked as required without cutting the whole plant.

Winter Density Suitable for autumn sowing, this dark-green variety is intermediate in appearance between cos and cabbage lettuce.

Valdor Suitable for autumn sowing (do not sow in spring or summer), this winter variety has large heads and tolerates cold, wet conditions.

Marrows and Courgettes

Although some varieties are suitable for both courgette and marrow production, F_1 hybrids, which are usually earlier and higher yielding, are better for courgettes. When grown as marrows, current F_1 hybrids have a tendency to reach a large size before attaining the thick skin that is desirable for storing marrows. In addition, the shape of current F_1 hybrids is not as cylindrical as that of the popular open-pollinated *Green Bush* types, which are best grown as marrows. *Emerald Cross* (now known as *Greyzini*) is an F_1 hybrid that produces high yields of mottled, light-green, club-shaped marrows but it is not particularly suitable for courgettes. The following varieties have done well in ADAS trials.

Open-pollinated bush type

Green Bush A heavy yielding, compact variety with cylindrical striped marrows that are sometimes variable in their colour pattern.

Green Bush – Smallpak* A shorter-fruited form of *Green Bush* that produces marrows which mature before attaining a large size.

Open-pollinated trailing type

Long Green Trailing A variety with long, cylindrical-shaped, striped fruit that sometimes thickens at the blossom end.

F_1 hybrid bush type

Burpee Golden Zucchini A good cropper with medium-long golden fruit suitable for courgettes.

Chefini* Produces large, dark fruit sometimes with light colouring on the underside. Young fruit gives cylindrical, dark-green, shiny courgettes.

Early Gem Also known as *Storr's Green*, a very early heavy yielder with mottled, dark-green fruits that are good as courgettes.

Zucchini A good cropper with mottled-green fruit suitable for use as courgettes.

Onions

The introduction of Japanese varieties of onions to the UK for autumn sowing has increased the versatility of the bulb onion crop. Although the bulb shape and skin quality of current varieties of the autumn-sown crop fall short of that of the spring-sown, the food values of the two types are no different, and the early harvest period (June/July) of the autumn-sown crop fills a niche in which home-grown bulb onions were not formerly possible. Whereas the autumn-sown crop is best grown by direct sowing, the spring-sown crop may either be raised by direct sowing and thinning, by the use of onion sets, or by plant raising in peat-blocks then transplanting.

The listed varieties for autumn-sown and spring-sown bulbs and for salad onions, are based mainly on results from annual trials by NIAB at several sites.

Bulb onions

Spring-sown

All the listed varieties have globe-shaped bulbs.

Early (mid-August harvest)

Rijnsburger-Adina* Although early, this variety has rather pale thin-skinned bulbs that are not suitable for long-term storage.

Early main crop (*late August harvest*)

Rijnsburger-Wijbo* This matures only a few days earlier than maincrop varieties and has bulbs of good skin quality but pale colour.

Main crop (*September harvest*)

Hygro A high-yielding, F_1 hybrid with globe-shaped bulbs which store well if kept dry at low temperature.

Rijnsburger- Balstora* A high yielder with dark, straw-coloured bulbs of very good skin quality; also stores very well.

Sets

Several varieties are sold as sets for planting in late March or April; they should not be planted earlier (see Chapter 6).

Ailsa Craig A very popular variety with large, globe-shaped bulbs with straw-coloured skins; the flesh is mild-flavoured.

Sturon A good-yielding variety with round bulbs that keep well.

Stuttgarter Giant A variety with large, flat onions that store well.

Pickling onions

Paris Silver Skin, Barletta and *The Queen* are all good silverskin onions suitable for pickling. It is also possible to use brown-skinned varieties normally grown for large bulbs by growing them at close spacing; *Giant Zittau* is suitable for pickling.

Autumn-sown (*not suitable for storing*)

Express Yellow O-X One of the earliest maturing (early June) hybrids, with pale, semi-flat, thin-skinned bulbs, it has lighter yields than later-maturing varieties.

Imai Early Yellow A mid-season (mid-June harvest) variety with high yields of flat-shaped bulbs with yellow skins of moderate thickness.

Senshyu Semi-Globe Yellow Matures about ten days later than *Imai Early Yellow*, it has high yields of globe-shaped bulbs with darker skins than other varieties.

Salad onions

White Lisbon The traditional salad onion used for pulling young in the spring. Quick growing with silvery skin and mild flavour.

White Lisbon-Winter Hardy Seed not as widely available as that of *White Lisbon*, this new variety is the hardiest of salad onions for harvesting in spring but tends to bulb faster than *White Lisbon*.

The traditional salad onion belongs to the species *Allium cepa* while so-called Welsh or Japanese bunching onions are *Allium fistulosum*. These bunching onions are usually sold under a Japanese name. *Hikari** is available from at least one supplier, and over the next few years it should be possible to identify other good varieties.

Parsley

Strictly speaking, parsley is a biennial herb, but because it is so popular and, unlike most herbs, there are a number of different varieties, it has been included here.

The varieties that are listed have done well in trials at the NVRS.

Bravour* A new variety with dark-green, curled foliage on longish stalks.

Moss Curled A reliable, dark-green variety of long standing with finely-curled foliage.

Paramount – Imperial Curled A vigorous variety of medium height with dark-green, curled foliage.

Parsnip

On soils where canker is a problem (usually on rich organic soils), canker-resistant varieties should be grown.

The listed varieties have done well in NIAB trials.

Resistant to canker

Avonresister Small bulbous, cream-fleshed roots, that are sweeter than other varieties; highly-resistant to canker.

White Gem Small- to medium-sized roots that are wedge-shaped to bulbous. Smooth skin with white flesh. Good resistance to canker.

Susceptible to canker

Offenham Medium to large, bulbous- to wedge-shaped roots; requires deep soils for best results.

Peas

The success of pea breeders in producing dwarf 12–36 in. (30–90 cm) high, modern varieties has made this crop much easier to grow. Heights will vary considerably, depending on soil fertility and growing conditions and a variety sold as '60–75 cm in height' may reach 36 in. (90 cm) under favourable circumstances.

Round-seeded varieties are much hardier than wrinkled (marrowfat) and are used for November or early spring sowings. Those varieties with wrinkled seeds should not be sown before the beginning of March, even in sheltered areas. Because of the higher sugar content and better flavour, wrinkled seeds are of better quality than the round-seeded varieties.

Varieties are usually categorized as first earlies, second earlies and main crop on the basis of the time that they take from sowing to reach maturity. As a guideline, early varieties usually take 11–12 weeks, second earlies from 12–14 weeks and main crop 14–16 weeks. Because of the overlap that may occur between the different maturity types, you may find occasionally that one seedsman lists a variety as a second early, while another advertises it as main crop. The time to maturity will vary to some extent with growing conditions. Maincrop varieties, even modern ones, tend to be taller than earlies.

Sugar or mangetout peas are eaten in the pod, just before the peas swell. The so-called asparagus or winged pea (it has four-winged pods) are not peas at all, although they belong to the same family (legumes) as peas. They too are eaten when young, and when cooked the young pods have a flavour that resembles that of asparagus. As the asparagus pea is extremely susceptible to frost it should only be sown when the danger from frosts is past.

Choosing a variety

Most of the official trials that are done with peas deal with the highly specialized vining pea crop for commercial growers. In addition, however, experience in growing different varieties by gardeners has identified a number of reliable varieties. The varieties that are listed therefore include those that have proved to be highly satisfactory over a number of years, as well as those that have done well in vining pea crop trials.

First earlies

Feltham First A round-seeded variety, height 18 in. (45 cm), suitable for early sowing, particularly under cloches; it has only moderate quality.

Early Onward A wrinkled pea, height 24 in. (60 cm), suitable for sowing from March onwards, this variety gives high yields of good quality and freezes well.

Hurst Beagle A wrinkle-seeded variety, height 18 in. (45 cm), suitable for sowing from March onwards, it produces good quality, sweet peas which freeze well.

Kelvedon Wonder A wrinkled pea, height 18 in. (45 cm), suitable for sowing after March, its tolerance to mildew makes it suitable for autumn-cropping when this disease is often bad.

Little Marvel One of the older, wrinkle-seeded varieties, height 16 in. (40 cm.), which does well under cloches, producing peas of very good flavour.

Meteor A hardy, round-seeded variety, height 18 in. (45 cm), for early sowing.

Second earlies

All varieties listed here have wrinkled seeds and should not be sown before March.

Hurst Green Shaft A very good variety, height 24–32 in. (60–80 cm), that gives high yields of good quality peas suitable for freezing.

Onward An established favourite, height 24–30 in. (60–75 cm), which gives high yields of good quality peas that are suitable for freezing.

Victory Freezer Synonym *Kelvedon Monarch*, height 18–24 in. (45–

60 cm), gives high yields of sweet peas suitable for freezing.

Main crop

The varieties listed here have wrinkled seeds and should not be sown before March.

Senator A tallish variety, height 36 in. (90 cm), with high yields of peas with good flavour.

Lord Chancellor Formerly known as *Chancelot*, this is a tall, later-maturing variety, height 36–48 in. (90–120 cm), with good yields.

Potatoes

'Quality' in a potato variety is difficult to define, as tuber appearance, dry matter content, flavour, texture and cooking characteristics all play a part. Also, varieties behave differently in different regions and personal and regional tastes differ. For example, among maincrop potatoes, *Kerr's Pink*, a floury potato, is popular in Scotland, whereas *Majestic*, a good keeper with a slight tendency to discolour after cooking but which makes excellent chips and sauté potatoes, has greater consumer appeal in England. Both these old varieties are being replaced by more modern ones, as are the first earlies *Home Guard* and *Arran Pilot*.

For the individual gardener it is advisable to grow a number of different varieties so as to determine which variety or varieties has greatest appeal. For those interested in older varieties with special qualities, D. MacLean of Dornock Farm, Crieff, Perthshire, publishes (a stamped envelope is required) a list of varieties 'with special properties and for special purposes'.

Early potatoes are usually grouped into first and second earlies. In sheltered, more southerly areas, first earlies can be ready for harvest in late May but generally speaking are ready for digging in June/July, while second earlies are available in July/August. With earlies it is best to use them when fresh and they are best served boiled, though larger ones make good chips and can also be used to make sauté potatoes.

Maincrop potatoes are usually harvested from September to

October, and can be stored over winter.

Potato-cyst eelworms (potato 'sickness') can be troublesome in gardens, particularly on land which has repeatedly grown potatoes. As there are no practical curative measures, varieties with resistance to any of the strains of this pathogen are at an advantage over susceptible varieties on infested soil.

The majority of varieties which are listed in order of maturity have done well in NIAB trials and should be readily available.

First earlies

Maris Bard A very early, high yielder with white, short, oval tubers with white flesh.

Ulster Sceptre Very early, this variety gives high yields under a wide range of conditions. The long, oval tubers have white skin and flesh. Seed should be handled carefully because of its susceptibility to gangrene.

Sutton's Foremost Produces a heavy crop of white-skinned, oval-shaped tubers that cook well and have a good flavour.

Pentland Javelin One of the later first earlies, this variety is slow to sprout but bulks quickly, giving high yields of uniform, roundish, white tubers with white flesh. Is resistant to one race of potato-cyst eelworm and can therefore yield well when susceptible varieties fail.

Second earlies

Estima* A very high-yielding variety with attractive oval, yellow-skinned tubers with pale-yellow flesh; it is susceptible to powdery scab. A Dutch variety increasing in popularity in commerce, seed should soon become more readily available.

Maris Peer A moderate yielder of uniform, medium to small, oval tubers with white flesh and skin; it is extremely susceptible to drought and prefers fertile, moist conditions.

Red Craigs Royal A red mutant-form of *Craigs Royal*, this variety bulks quickly and gives a moderate yield of oval, white-fleshed tubers that cook well.

Wilja A high-yielding Dutch variety with very uniform, long-oval yellow tubers of uniform size that have a pale-yellow flesh.

Main crop

Desiree An early maincrop with high yields of red-skinned, attractive long-oval tubers with light yellow flesh of good cooking quality with little disintegration on boiling. Scab and slugs can be troublesome.

King Edward Gives good yields of oval, white-skinned tubers with red splashes. Yields vary with conditions as it is very susceptible to drought. The cream-fleshed tubers are of very good cooking qualty. *Red King Edward* is a fully red-skinned variant.

Maris Piper A consistently high-yielding early-maincrop variety with attractive, medium-sized tubers that are short-oval, have white skin and cream flesh, and are of good cooking quality. Although it is susceptible to drought, common scab and slug damage, it is resistant to one strain of potato-cyst nematode.

Pentland Crown A very high-yielding, late variety with large, short-oval tubers with white skin and flesh. It is very resistant to common scab and to drought but does not always store well.

Pentland Dell A high-yielding early maincrop which produces numerous, uniform and attractively-shaped white tubers of excellent quality apart from slight after-cooking blackening. Well-sprouted seed should be used for planting which should be delayed until the soil has warmed up.

Pentland Squire A very high-yielding early maincrop variety with large, attractively-shaped white tubers. Closer spacing will help to reduce tuber size, as large tubers can develop hollow hearts. Drought-resistant, the tubers have excellent cooking quality and store well.

Radish

This quick-growing salad crop has varieties with a wide range of shapes and colours. Japanese varieties, usually with long white roots, are likely to become available over the next few years.

The varieties recommended here have done well in trials at the NVRS.

Summer

Cherry Belle A popular red-globe variety with crisp, white flesh.

French Breakfast A popular variety with red, cylindrical roots with a white tip.

Long White Icicle A long-rooted, white-skinned variety with a pleasant, mild flavour.

Saxa – Red Prince* An early red-globe variety with white flesh.

Saxerre* A new quick-maturing, red-globe variety suitable for sowing in cold frames or under cloches.

Winter

Black Spanish Round A globe-shaped winter variety with a black skin and white flesh.

China Rose A rose-coloured, cylindrical variety with white flesh.

Rhubarb

Although seed of some varieties can be purchased, it is advisable to purchase rhubarb crowns to establish this crop. Such crowns should be available in garden centres and from local specialist rhubarb growers.

A number of varieties have done well in trials at Stockbridge House EHS, Yorkshire, and include the early variety *Timperley Early*, which is suitable for early forcing. *Victoria* and *Hawkes Champagne* are good mid-season types while *Cawood Delight** is a high quality, red-fleshed, late variety that is not readily available.

Spinach

Although an easy crop to grow, if a spinach variety is sown at the wrong time of year, plants will bolt to seed, particularly in dry weather. Although some varieties are described as prickly, the seeds, and not the leaves, are prickly.

Few official trials have been done in this crop, so that the listed varieties are those generally reported to have given good results over a number of years.

Broad-leaved Prickly A very winter-hardy variety with thick, fleshy, dark-green leaves.

Viking* A variety with dark-green leaves suitable for sowing from March to July.

Swedes

Swedes are hardier and usually larger than turnips with a flavour that many prefer to turnips. The availability of varieties resistant to mildew is helping to increase the popularity of this vegetable.

A variety that has done well in NIAB trials is

Marian A high-yielding variety with good resistance to clubroot and to powdery mildew. It is globe-shaped with a purple skin and has attractive yellow flesh with a good flavour.

Sweet corn

Although still a marginal crop in the UK, the breeding of earlier-maturing F_1 hybrid varieties makes it possible for gardeners in the southern half of England to obtain reasonable success with 'corn on the cob'. The F_1 varieties recommended here should mature about 120–130 days from sowing.

The listed varieties have done well in trials done by the NIAB.

Earliking Good yield, moderate grain length.

Kelvedon Sweetheart* High yield, well-filled grains on most of cob.

Northern Belle Very high yield of well-filled grains on most of the cob.

Tomatoes, outdoor bush or dwarf

Most tomatoes are grown in greenhouses or under cloches and usually require staking, stopping by pinching-out the growing tip once three or four trusses have formed, and pinching-out of side-shoots. The breeding of bush, or determinate, varieties that require no staking or pinching, together with new methods for raising seedlings (see Chapter 2 of *Know and Grow Vegetables*), has made possible the growing of true 'outdoor' tomatoes, particularly in the south. Although these determinate varieties require no stopping or staking, black plastic can be used to keep fruit off the ground and helps control weeds; straw bedding may help to spread disease.

Results from trials done by NVRS show that good results

should be obtained from the following varieties:

Sleaford Abundance An F_1 hybrid bush variety, with little leaf and a heavy crop of small- to medium-sized fruit of good quality and shape. Has a sprawling habit.

Alfresco* A more vigorous, new F_1 hybrid bush variety from the breeder of *Sleaford Abundance*; about 7 days later maturing than *Sleaford Abundance*.

Sigmabush* An open-habit F_1 hybrid with medium-sized fruits; similar maturity time to *Sleaford Abundance*.

The variety *The Amateur*, which is probably one of the bush varieties best known by gardeners, has been outyielded by *Sleaford Abundance* and *Alfresco* in NVRS trials.

Turnips

A fast growing brassica crop, turnip roots are of many shapes, sizes and colours. The white-fleshed varieties are best used when young, otherwise they become hot and stringy.

The varieties listed are reported to have given good results over many years.

Golden Ball A yellow globe-variety with yellow flesh. Fairly hardy, it also keeps well.

Manchester Market Green Top-stone A green-topped, white globe-variety with a mild flavour. Fairly hardy, it also keeps well.

Purple Top Milan A very early variety with flat-shaped roots; white with purple top.

Snowball A very early, white globe-variety with mild flavoured flesh.

2 Planning continuity of supply

Most gardeners aim to produce a continuous supply of good quality vegetables for the kitchen for as long a period as possible, but even the expert has to put up with either gluts or shortages at times during the year. Of course with crops like peas, beans, tomatoes and beetroot, a glut of produce can be preserved by freezing, canning, bottling or pickling, but not everyone is in a position to make use of surplus produce in this way. Furthermore, with salad crops such as lettuce even the practice of sowing seed every few days to maintain a succession can lead to two or three sowings being ready together. The surplus cannot easily be disposed of or preserved and often finds its way onto the compost heap. Successional sowings can have many pitfalls and the aim of producing a regular supply of fresh quality vegetables is far easier stated than achieved.

Problems of irregular supply can be directly attributable to the unpredictability of the weather. Its variability results in seeds germinating sooner, or later, than expected depending on the soil temperature and moisture content. The seeds may germinate but fail to push up through a hard 'cap' on the soil surface which may be brought about by heavy rain breaking down the soil crumbs and cementing the surface layer. Even when the seedlings have finally emerged, they may experience weather which is fine and warm, enabling the plants to grow very rapidly, or cold, dry and windy which will seriously check growth and retard development.

This unpredictability of the weather and its effects makes the faint-hearted accept fatalistically that nothing can be done to improve the situation. On the contrary, it will be shown that with a knowledge of the way in which weather factors such as temperature affect the growth and development of different vegetable plants, (see Chapter 6), the effects of varying weather-conditions can be minimized considerably. The problems of getting predictable supplies of each of the common vegetable crops will then be discussed in some detail in relation to recent research results. Finally, proposals are given for sowing and planting programmes including the production of earlier and later crops by protection and other cultural methods, and for extending the season by storage. Detailed information on the storage of individual vegetable crops is given in Chapter 3.

The choice of crops, the proportion grown of each, and their season for harvesting are matters decided according to individual taste and opportunity (which will include the size and physical advantages and disadvantages of each vegetable plot). Similarly, the productivity of a plot measured in terms of the total weight of vegetables produced per unit area in a year will also depend to a large extent on the amount of time and effort put in by the individual gardener, and also the intensity of cropping practised through growing two or three different crops on the same piece of ground simultaneously, or in succession. These particular aspects will not be dealt with here as there are several sources of reliable information to guide the reader, such as that produced by the RHS in *The Vegetable Garden Displayed*. Rather, the aim of this chapter is to outline principles and the different ways in which supplies of vegetables can be maintained over long periods by minimizing as far as possible the unpredictable effects of the weather.

WEATHER AND THE GROWTH OF VEGETABLES

The growth of all plants is greatly affected by weather conditions and especially by temperature, sunshine and rainfall. These three factors in particular will not only affect the *growth rate* of plants but consequently the yield and the time of maturity. In addition, temperature can influence the *development* of certain vegetable crops especially those whose buds, fruits or seeds are harvested.

Above a certain minimum temperature, which varies for different crops, seeds will germinate and seedlings and plants will grow rapidly if the temperature is, say, in the range of 65–75°F (18–24°C) providing moisture supplies are adequate. Under these conditions the plants will develop more or less normally and will produce *sooner or later* the root, shoot, leaf, bud, flower, pod, fruit or seed – whichever part of the plant we actually eat. 'Sooner or later' is the operative phrase, for it is not generally realized that to maximize the growth and development of a particular vegetable plant, the optimum conditions of temperature, radiation and rainfall may change with the different phases of development. Obviously, problems can be caused if adverse weather conditions coincide with a particularly sensitive stage of growth.

For example, the seeds of butterhead varieties of lettuce will not germinate if the soil temperature exceeds 77°F (25°C) – a temperature which is frequently reached in the surface layers of soil during hot periods in summer. The seeds experiencing these temperatures become 'thermo-dormant' and will not germinate until the soil becomes cooler and even then only after some delay. So with this crop a planned successional sowing programme may fail because in spells of hot weather seed germination will not take place on time, even when the soil is moist, and gaps in the succession of crops will inevitably result. This problem has been discussed in some detail in our previous book (Chapter 2) and readers will know that there are ways of overcoming it which will be briefly described again later.

In general, assuming that this difficulty of thermo-dormancy is overcome, with a vegetative crop such as lettuce where we are only interested in leaf growth, the higher the temperature the faster the leaves grow. As a rule, the growth rate of plants doubles with every 10°C rise in temperature, hence the spurt in growth during warm periods in spring.

For many crops, however, the plants' response to the weather is not quite so straightforward, for special combinations of environmental conditions may be essential at particular stages of growth in order to 'switch' their pattern of growth to produce flowers, pods or bulbs (see Chapter 6). For example, the majority of cauliflower varieties require periods of relatively cool temperature below 70°F (21°C) to 'trigger off' the formation of the cauliflower curd. After the plants have passed through their early stage of puberty, when they are *developmentally* unreceptive to temperature, the accumulated effects of lower temperatures cause the plant to stop producing leaves and instead it initiates the curd. By a process of evolution, and more recently of breeding, we have evolved a series of cauliflower varieties each of which requires a different amount of exposure to relatively low temperatures before the plants will form curds. As a result the earliest varieties of summer cauliflowers, such as the Snowball type, need little in the way of cold stimulus to form curds (and may frequently produce premature 'buttons'). Progressively later-maturing varieties require more and more exposure to low temperature and some of the winter-maturing group may require the equivalent of twelve weeks at a temperature less than 60°F (15°C).

So, depending on the prevailing temperatures during the growth of a crop, the curds of a variety will be formed earlier (if the weather is cooler) or later (if warmer) than average, and so the time the curd is ready for cutting will be affected in the same way. We will see later how the different 'cold' requirements of the various varieties can be used to advantage in ensuring continuity.

More information on the precise effects of weather factors on the growth and development of vegetables is given in Chapter 6, whilst the effects of rainfall and watering have been described in Chapter 4 of the companion volume *Know and Grow Vegetables*.

HOW CAN PLANNED PRODUCTION BE MADE MORE RELIABLE?

Whether we want to produce a reliable supply of a vegetable over as long a period as possible to supply the kitchen, or whether one particular crop has to be timed precisely for entry at a local show, the nature of the problem is the same: how can we minimize, or even eliminate, the unpredictable effects of weather, especially those of temperature on our crops' growth.

There are five main approaches we can adopt depending on the type of crop and the degree of reliability required; if necessary two or more can be combined in practice:

- by the use of average dates for sowing and planting,
- by taking into account prevailing temperatures in successional sowings,
- by sowing different varieties at the same time to obtain a succession,
- by modifying the environment around the plants, and
- by developing new ways of growing crops.

Use average dates for sowing and planting

For certain vegetables calendar dates can be a useful guide for timing sowing or transplanting. Although weather conditions can be very different in, for example, the first week of April from year to year, the timing of the harvest of some crops is not critical, or, with others, can be very little altered by the time of sowing. For example the roots of crops such as carrot, parsnip or beetroot may get larger the longer they are left in the ground but do not, in general, deteriorate in quality. Within reason, therefore, sowings can be made at fairly wide intervals of time and the plants can be pulled as required. These crops can also be stored for long periods after growth has finished.

Other crops, such as certain varieties of cabbage, can remain in a mature state for up to two months without being harvested and without deterioration, whilst runner beans can provide a succession of beans from the same plant.

Yet other crops such as onion form bulbs and ripen predictably year after year because bulb formation and maturity are controlled almost entirely by the *length* of the day (see Chapter 6); small variations in the time of sowing may modify the bulb size rather than the time when it ripens. Furthermore the bulbs can be stored for many months.

Thus with all these types of crop there are no great problems in maintaining supplies over long periods of time and there is not the same need to time sowings precisely to obtain continuity of supply.

With other crops such as lettuce, which have a short harvest period and do not keep, individual sowings are often made on predetermined dates to provide a succession. The gardener soon realizes from experience that the intervals between sowings need to be longer in the spring than in the summer. This is because as the average daily temperature rises throughout the spring to a maximum in mid-summer and then falls in the late summer and autumn (see Fig. 2.1), the time to produce a hearted lettuce will become progressively shorter with later

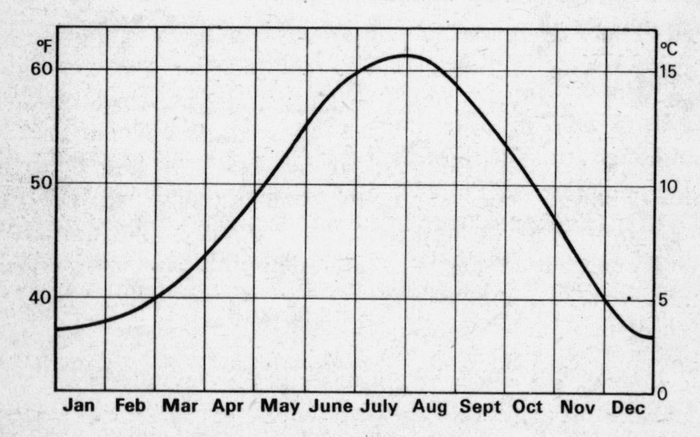

Fig. 2.1 Seasonal trend in mean daily temperature at Wellesbourne, Warwickshire averaged over the years 1952–1979.

sowings from April until June. Obviously, then, the intervals between sowings need to take this seasonal trend into account and unequal time intervals can be quite successful in regulating the time of cutting of successive crops.

Make successive sowings dependent on prevailing temperatures

With crops such as peas and French beans a continuous supply of fresh produce can be obtained by making successional sowings at intervals. The problem, as with that of lettuce just described, is to decide on the length of the intervals between sowings. If sowings are too frequent then crops from successive sowings may mature together; if the intervals are too long then there will be temporary periods of shortage. Fortunately a system has been devised for processing crops such as peas.

The basis of the method is that, as plant growth rates are related to temperature, the intervals between successive sowings should be varied depending on the *actual* temperatures since the last sowing, rather than intervals based on average seasonal trends (see Fig. 2.1).

How can this be put into practice?

The unit of measurement used to combine the temperature with its duration is the 'day-degree', 1 day-degree being the equivalent of a rise of 1°F (or 1°C depending on the scale being used) from the base temperature for growth for a period of 24 hours. For most crops the assumption is made that plant growth will start when the temperature reaches 42°F (5.6°C), although for some crops such as French beans and sweet corn the minimum is 50°F (10°C), and for cucumbers as high as 58°F (14°C). Growth will then be proportional to the number of day-degrees accumulated over a period of time. The gardener can calculate these accumulated day-degrees approximately as follows:

(1) Record each day at 0900 GMT the maximum and minimum temperatures shown on a 'max and min' thermometer (Fig. 2.2); let us take as an example, a maximum of 69°F and a minimum of 43°F.

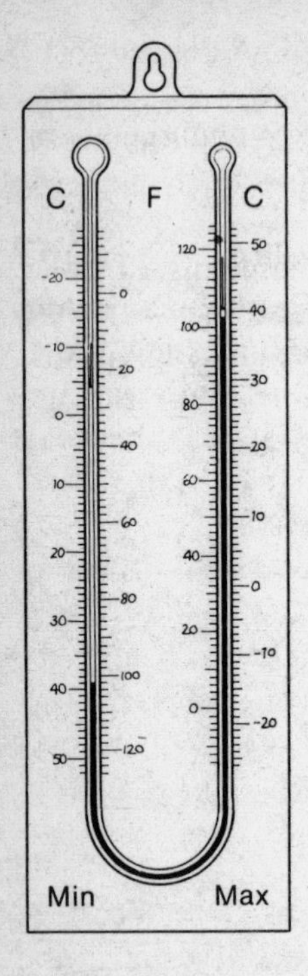

Fig. 2.2. Maximum-minimum thermometer suitable for garden use.

(2) Calculate the average daily temperature by adding the maximum and minimum temperatures together and dividing by 2, thus 69 + 43 = 112 ÷ 2 = 56.

(3) Subtract the base temperature for growth (42°F). This will give the number of day-degrees for that day, in this example 14, to which growth is related. Running totals of accumulated day-degrees can then be recorded from any starting date using either a Fahrenheit or Celsius scale.

How can this method be used to plan a sequence of successional sowings of a crop? If we know the average number of accumulated day-degrees required for a given variety of vegetables to reach maturity from sowing, and if we also know the long-term seasonal trends of day-degrees for our particular site (see Table 2.1), we can predict fairly accurately when the crop from a particular sowing will be ready to pick. We then decide how rapidly we want our successional crops to follow each other in time of maturity. For example, we may want them to be ready at weekly intervals in July, which, at Wellesbourne (Table 2.1) is equivalent to intervals of about 130 day-degrees F. The individual sowings in May will then have to be separated by intervals of 130 *actual* day-degrees accumulated from the previous sowing, rather than by a number of days. In this way we take into account the seasonal trends and vagaries of the weather and a more reliable succession of crops can be obtained.

Table 2.1 Average weekly totals of accumulated day-degrees above 42°F at Wellesbourne, Warwickshire

Week number	*Jan*	*Feb*	*Mch*	*Apl*	*May*	*June*	*Jly*	*Aug*	*Sept*	*Oct*	*Nov*	*Dec*
1	6	7	12	26	48	101	122	127	109	81	28	13
2	6	7	15	30	54	106	124	125	100	71	23	13
3	6	8	18	41	62	109	128	125	90	54	19	10
4	6	10	21	44	72	114	126	122	87	52	16	10
5		10			79			115		42		
Totals	24	42	66	141	315	430	500	614	386	300	86	46
Running totals	24	66	132	273	588	1018	1518	2132	2518	2818	2904	2950

At present this method is successfully used for programming commercial production of peas, French beans, sweet corn and outdoor tomatoes, and experimentally for predicting the harvest times of lettuce, carrot and mini-cauliflower. Because the necessary information is not yet available it cannot, at present, be used for other vegetable crops. However, the enterprising gardener is recommended to experiment with the method on a range of crops in order to collect quantitative information for his site which will become increasingly valuable to him when planning his vegetable programme in future years.

Use different varieties

For some vegetables a simple method of getting a continuous succession of produce is to grow a number of varieties which take different lengths of time to reach maturity from the same sowing date. We have already seen that different cauliflower varieties have different 'cold' requirements before curds are formed, and this affects the times when the heads are ready for cutting. Different varieties of other vegetables also respond differently, but consistently, *relative to each other*, to various aspects of the weather, such as temperature thresholds for cold tolerance in beans and day-length effects on bulbing in onions.

The subject has important implications for planning continuity of production. For example, if a number of varieties which take different lengths of time to mature are taken of, say, Brussels sprouts, French beans, cauliflower, and onion and are sown on a single date over a number of years, the varieties of each crop will always mature in the same relative order despite the different weather conditions in each year. The relative order of maturity of varieties used for commercial production has been shown to be so consistent by the NIAB in their official trials (see Chapter 1) that leaflets specify the average order of earliness (or lateness) in numbers of days of the best varieties of the major vegetable crops.

Many of the garden varieties, however, have not been included in this type of trial because they do not have the necessary characteristics essential for today's commercial production systems. Thus, there may only be a more general type of information on the relative maturity of such varieties obtainable from seedsmen and their catalogues. However, for the gardener, general descriptions of 'early' or 'mid-season' could be sufficiently precise as there is often sufficient variability between individual plants of the same variety to supply the kitchen until the next variety is ready.

So depending on the type of crop and the length of the period over which fresh produce is wanted, as few as 2 or 3, or as many as 6–10 different varieties may be needed in order to ensure continuity. Here, perhaps, lies the main disadvantage of this method for the gardener – the cost of buying more packets of seed than normal. But unused seeds can be stored from one year to the next under the correct conditions (see

Chapter 2 of *Know and Grow Vegetables*), or fewer varieties can be grown with more times of sowing.

This general approach of using varieties to provide continuity has a lot to commend it, not the least being the simplicity of putting it into practice. It also has the advantage that it is easier to choose the time when seedbed conditions are ideal for sowing because the timing of seed sowing is not so critical as with some of the other methods previously described.

Modify the plants' environment

When the various aspects of the environment are precisely controlled in air-conditioned glasshouses, 'growing-rooms' and cabinets for experimental purposes, the growth, development and time of maturity of vegetable plants can be predicted with complete accuracy. Obviously this is not a practical solution for the outdoor vegetable gardener, but it does show that the more we can minimize adverse effects of the weather the earlier and more predictable the maturity of our crops will be.

The main problem to overcome, both in the early spring and for extending the season in the autumn, is that of low temperatures which reduce growth and may even kill plants if they are at sensitive stages of growth. Drought conditions or exposed areas are more easily remedied by watering or providing shelter.

The adverse effects of low temperature on crops can be minimized in several ways: directly, by covering the seeds or plants with cloches, polytunnels, various mulches, or even with newspapers, for different lengths of time; or, indirectly, by raising plants in greenhouses or frames before planting them out into the open later in the season when the effects of cold spells are not so severe.

Glass cloches of the traditional 'bell', 'barn' or 'tent' type have been used for many years to produce early or late crops of vegetables by providing a warmer environment with protection from wind, storms, and frost. In recent years new materials have tended to replace glass in the traditional cloche, and tunnels made of clear thin polythene supported by wire hoops are a more recent cheap alternative method of protecting crops. Both soil and air temperatures are warmer under cloches and tunnels because the loss of radiant heat from the soil is reduced. If crops such as lettuce are grown throughout their life under such forms of protection they can be ready for cutting up to 3–4 weeks earlier. However crops such as beans, peas and tomatoes are only started under cloches and tunnels to get earlier germination and plant establishment, which usually results in earlier maturity. Other crops can be protected in the autumn to hasten ripening, or extend the season of production.

Mulches laid over the soil surface can hasten seed germination and early seedling growth by making the soil warmer. Again in recent years traditional materials such as straw, leaves and peat are being replaced by different forms such as paper, paper bonded with polythene, and other forms of polythene sheet. Clear or black polythene is preferred commercially. Rapid developments are taking place commercially on mulches; biodegradable polythene is now available with varying lengths of life, whilst a slitted polythene sheet which is laid down as a mulch is forced up by the growing plants to form a tunnel partially ventilated through the slits or holes in the polythene.

The most important way of modifying the effects of the weather on plant growth and development is, of course, to raise plants for the first few weeks of their life under good conditions in greenhouses and frames before transplanting them into the garden. The advantages of this method compared with sowing seeds directly in the garden may be summarized as follows: it allows earlier crops to be produced; it enables crops such as sweet corn and tomatoes to be grown more reliably in our short summer season; it provides better conditions for young plants in the early stages of growth which are often critical if good crops are to be obtained; it enables more crops to be grown because the plants occupy the ground for a shorter period of time; and, not the least of the advantages, raising plants under better conditions usually results in more plants being raised from a given quantity of seed. Nearly all of the brassicas are better transplanted into the garden as well as salad crops such as lettuce, bulb onions and leeks, and the plants can be raised in different types of pots, peat or soil blocks, tubes or boxes. Plants grown individually in containers or blocks establish themselves after transplanting much quicker than those bare-root transplants which have had to be separated from each other before planting; there is, therefore, less check to growth and earlier crops are obtained from them. In general the younger the plants are transplanted the quicker they become established and the better are the resulting crops, provided they have been hardened off and provided also that the soil and weather conditions in the garden are reasonable. This subject of transplanting and plant raising has been dealt with in detail in Chapter 2 of *Know and Grow Vegetables*.

Use other cultural methods

All of the common cultural practices used in the garden will affect the way a crop grows and the time it is ready to harvest. However, there are specific cultural methods which can help to produce earlier or later crops of some of our common vegetables, as a few examples will show.

Early crops of broad beans, cauliflower, and salad and bulb onions can be regularly obtained, but not guaranteed, by making autumn sowings of certain varieties which will usually overwinter in an average year without too many losses. The plants have to reach a certain size by late autumn otherwise they will not survive the winter weather and therefore the sowing time for these crops is critical. Detailed information will be found later in this chapter under each crop.

With crops such as parsnip and parsley which are sown directly into soil in the early spring, and take several weeks before seed germination and the seedling emerges, pre-germination of the seed by sowing under optimum conditions in sandwich boxes indoors (see Chapter 2 of *Know and Grow Vegetables*) before fluid-sowing in the garden in March, can give 2–3 weeks earlier seedling emergence and much earlier crops.

With certain crops, 'stopping' of the plant by pinching-out the growing point will encourage earlier development of the upper 'buttons' of Brussels sprouts, of pods on beans, and fruit on tomatoes, whilst removing the leaves shading the fruit trusses of tomatoes will hasten the ripening of the fruit if the weather is warm and sunny.

In a somewhat similar way the use of transplants in brassica crops such as cauliflower provides a means of regulating the supply of mature cauliflower curds. If, from one sowing in an outdoor seedbed, a few plants are lifted at weekly intervals from, say, 4–9 weeks old from sowing, a succession of mature heads can be produced for the kitchen, the older transplants maturing later than those planted earlier.

Some of the more recently developed systems of crop production have considerably modified the maturity characteristics and season of production of vegetables. For example, overwintered crops of bulb onions have been grown for many decades in this country, but with the varieties used even the earliest crops were not obtained until late July. With the introduction of the winter-hardy, intermediate day-length, bolting-resistant Japanese varieties, mature bulbs can be lifted at the end of May, six weeks earlier than even set-raised crops (see Chapter 6 p. 177). A very different example is the 'new' leaf lettuce, so-called because the plants are grown so closely-spaced that they do not produce a heart. The leaves are ready for harvest much earlier than the conventionally-grown hearted crop and the equivalent of four or five normal-hearted lettuce per week from mid-May to mid-October can be obtained from as little as 5–6 square yards (4–5 square metres) of garden using a single variety.

On the other hand, when plants of certain varieties of cauliflower are grown closely-spaced they produce small curds or 'mini-cauliflowers' which tend to mature all together and are ideal for home-freezing. The maturity time of these varieties, when grown in this way, is predictable throughout the season from July until October to within three or four days and so, if required, a daily supply of mini-cauliflowers can be reliably planned (see Fig. 2.3).

THE ROLE OF STORAGE TO EXTEND THE SEASON OF SUPPLY

Having done all that can be done to produce vegetables fresh from the garden for as long as possible, we are still able to extend the season of supply to a greater or lesser extent by various methods of storage. These range from the simple, such as putting lettuce heads into plastic bags before placing them in a refrigerator or other cool place to keep for a few hours or days, to the equally straightforward but rather more time-consuming methods used to prepare crops such as onions and beetroot for storing up to 6–8 months. The various alternative methods which can be adopted for the major vegetable crops are described in Chapter 3.

PLANNING FOR INDIVIDUAL CROPS

The following suggested schemes for planning continuity of supply should only be regarded as guidelines, for each family will have their own preferences and each garden will have its own microclimate resulting from the combined influences of location, aspect, soil, shelter and rainfall. Furthermore, sowing and planting dates must necessarily vary in different parts of the country, and common sense will dictate when adjustments will have to be made. Nevertheless, they will provide a basis on which to develop your own system for planning the timings of each crop. If detailed records are kept in a gardening diary you will find that they will be a great help in perfecting your programme in future years.

Broad beans

This is usually regarded as a short-season crop but beans can be produced in succession from the end of May until October. The pods must be picked young (before the scar on the bean turns black) if tender beans with the best flavour are wanted. The beans will not 'hold' on the plant without becoming tough, neither can they be stored fresh for any length of time. This means that a succession of crops needs to be grown if the picking period is to last more than 2–3 weeks. Fortunately, this is easily achieved by sowing a number of different varieties, or a single variety, on a number of occasions, as outlined in Table 2.2. The timings of the sowings at 2–3 week intervals are not critical and do not warrant the use of the day-degree method; they will depend more on the suitability of soil conditions for sowing. The results from unprotected sowings in the autumn are not completely predictable.

Table 2.2 Sowing sequences for broad, French and runner beans

Method of growing	*Varieties*	*Harvest period*
Broad bean		
Sequence A		
(1) Sow outside in autumn late Oct.–Nov.	Aquadulce	early June–July
(2) Sow under cloches Jan.–March (or in frames and transplant in April)	Longpod type e.g. Exhibition Longpod	late May–June
(3) Sow outdoors at 2–3 week intervals March–May	Longpod followed by Windsor varieties	late June–September
or Sequence B		
Sow a single variety successionally as above until July. Autumn sowings should be made under cloches	The Sutton (a dwarf variety)	late May–late September
French bean		
(1a) Sow under glass mid-April, harden off and plant early June	The Prince Tendergreen	late June–July
or		
(1b) Sow under cloches mid-April and protect until June	The Prince Tendergreen	early July–August

Table 2.2 Sowing sequences for broad, French and runner beans (Cont'd)

	Method of growing	*Varieties*	*Harvest period*
(2)	Sow outdoors mid-May and mid-June	Glamis Loch Ness Tendergreen	late July–September
(3)	Sow outdoors mid-July and cover plants with cloches September	The Prince	mid-September–October
Runner bean			
(1a)	Sow under glass mid-April, harden off and plant early June. Pinch out growing point	Kelvedon Marvel Sunset	end June onwards
or			
(1b)	Sow under cloches late April and protect until June	Hammonds-Dwarf Scarlet	early July onwards
(2)	Sow outdoors end of May	Achievement Enorma Prizewinner Streamline	early August–October
(3)	A late sowing in June in mild areas	as (2)	September onwards

French beans

The flavour of the runner bean is usually preferred to that of the French bean if there is a choice, and as a consequence French beans are often grown only for the early crops before runner beans are ready. The very earliest crops to be picked in June are grown throughout in a warm greenhouse. However, by successional sowings at monthly intervals from March or early April to mid-July, French beans can be picked from the garden continuously from late June until October (see Table 2.2). A minimum soil temperature of 50°F(10°C) is required before the seed of this crop will germinate. So, for the earliest outdoor crops seed should be sown in pots or blocks and the plants planted out in early June after being hardened off. Alternatively, seeds are sown under cloches in March or early April and the plants protected until the risk of frost is passed. Unprotected crops are sown from mid-May until mid-July, but for the latest sowing the plants will need to be covered with cloches when the first frosts are forecast in September or October.

All beans should be picked when young to encourage more to form and to prevent them from becoming tough and stringy. Many of the newer varieties such as Loch Ness and Tendergreen are stringless and pencil-podded, and are excellent for freezing.

Runner beans

By making two or three sowings, runner beans can be produced from the end of June until the first frosts kill the plants in the autumn. Many gardeners, however, do not try to get very early crops but make just one sowing of a single variety in the open ground in the latter half of May for picking from early August until September or October. Earlier crops can be obtained by mulching with a clear polythene sheet. Experiments have shown that the mulch can raise the soil temperature by 3°C which has resulted in earlier seedling emergence, flowering, and up to two weeks earlier picking.

Earlier crops can be obtained by sowing dwarf varieties such as Hammonds-Dwarf Scarlet in late April under cloches or tunnels protecting the plants until mid-June; they will then produce beans in early July. Even earlier crops can be produced by sowing suitable early-flowering varieties into pots of compost in April for raising under glass. These plants must be gradually hardened off before planting out in early June, and the growing points should be pinched out to encourage early flowering and fruiting. It should be remembered that these plants may need to be protected even after planting out if frost is forecast, because runner beans are very susceptible to cold and young plants will be killed by even a slight frost.

Plants from the main sowings will continue to produce flowers over a long period and these will set normally provided all the pods are picked young. Pods should never be allowed to mature on the plants (unless you are saving seed) and a plentiful supply of water should be given to the roots after flowering has started. Experimental work has shown conclusively that syringing of the plants with water does not improve pod-set and is a waste of time.

If as long a picking period as possible is needed, the sowing or planting sequence in Table 2.2 should be followed, making due allowance for the earliness of your own district. The beans cannot be stored fresh for any length of time but can be frozen or salted to extend the season.

Beetroot

It is not difficult to produce a continuous supply of beetroot from early June until November but the aim should be to get a succession of young tender roots by sequential sowings of globe varieties, the timings of which are not critical. The earliest crops of small beet are obtained from sowing bolting-resistant varieties such as Boltardy or Avonearly under cloches in March; transplanting of beetroot plants is not recommended unless the seedlings are raised in a container or blocks and planted when small. These crops can be followed by sowing the same varieties outside in late March or early April. Later sowings can be made at monthly intervals until July, growing these and other globe varieties or the long-rooted Cheltenham Greentop. For a late autumn crop, Detroit Little Ball is reported to be good for the last sowing in July in mild areas.

To produce beetroot for storage, globe varieties or Cheltenham Greentop are sown in late May or June and lifted in October.

Broccoli, sprouting

From a single sowing of three or four varieties, sprouting broccoli can be picked from January until May in most years. The single sowing is made in a seedbed outdoors between mid-April and mid-May and the plants are transplanted in June or July. If the young shoots are picked every few days when young and tender the plants will re-sprout and continue to produce for 6–8 weeks. The period of picking can be extended by growing as well as the variety purple-sprouting, early and late selections of it. The later-maturing selections are very hardy and will survive most winters. There are also white sprouting forms for cutting during March and April. The green type, known as calabrese, is discussed on page 63.

Brussels sprouts

Many new varieties have been introduced in recent years and it is now possible to produce sprouts from August until late March. This crop is a good example of continuity of production being achieved over a six-month period from October to March from a single sowing by a suitable choice of varieties, and five varieties have been listed by the NIAB for use by gardeners. Details are given in Table 2.3. The single sowing is made between mid-March and mid-April under cloches, in cold frames, or in a sheltered site in the garden, and the plants are transplanted in mid-May to early June at a close spacing of 24 in. (61 cm) square.

When very early crops are wanted in August and September an early variety such as Peer Gynt is sown under cold glass in late February or early March, and the plants transplanted in late April or early May.

It is possible to obtain continuity from October until early March by growing just one early and one late variety which hold their sprouts on the stem in good condition over a long period of time. Suitable varieties are Peer Gynt or Cor (Valiant) for picking before Christmas and Rampart for later production.

The buttons develop from the base of the stem upwards and are picked over several times when they have reached the required size. The removal of the growing point of the stem, 'stopping', will hasten the growth of the sprouts making them ready for picking earlier than from unstopped plants, but will restrict the length of the picking season. Stopping is best done when the lowest sprouts on the stem are at, or approaching, ½ in. (12 mm) in size. This practice is only useful for early crop-

Table 2.3 Sowing and varietal sequence for Brussels sprouts

Method of growing	*Varieties*	*Harvest period*
(1) Sow in cold frames in late Feb. and transplant late April or early May	Peer Gynt*	August–September
(2) Sow under cloches or in frames mid-March to mid-April and transplant mid-May to early June	Peer Gynt	September–October
	Valiant	October–November
	Perfect Line or Achilles	November–December
	Rampart	December–January
	Fortress	January–March

* 'stop' plants for early August picking (see p. 60)

ping, for it has little effect if it is done after the beginning of October. Apart from making the sprouts ready earlier, it tends to make all the buttons up the stem more uniform in size which may be an advantage if they are all to be picked at one time for home-freezing.

Cabbage

By a combination of suitable varieties and sowing dates it is possible to cut cabbages throughout the year. Some of the recently introduced varieties remain in good condition for 2–3 months after they are ready for cutting and this helps to maintain continuity of supply. There are pointed-, round- and drumhead types, and all are easy to grow. Sowing times and varieties can be conveniently grouped according to the season of production and a summarized continuity programme is given in Table 2.4.

For information on Chinese cabbage see p. 68.

Spring cabbage: A single sowing is made in late July or early August directly into the ground in which the plants will mature, or in a seedbed for transplanting in September. The plants are grown closely together in the row if they are for cutting as unhearted greens. The varieties used must be resistant to bolting, and suitable ones for April cutting are Durham Early and Avon Crest. If a wider spacing is used to allow the plants to heart, the varieties Harbinger and Avon Crest mature in May.

Table 2.4 Sowing and varietal sequence for all-the-year cabbage production

Method of growing	*Varieties*	*Harvest period*
(1) Sow directly into ground late July *or* seedbed and transplant in Sept.	Harbinger, Avon Crest, Offenham selections	March–May
(2) Sow in heat in late Feb. in blocks or pots, transplant in mid-April	Hispi Marner Allfruh	May–June
(3) Sow in cold frames in late March and transplant late May	Hispi, Marner Allfruh, Stonehead, Market Topper, and Minicole (in order of maturity)	July–September
(4) Sow outside early May and transplant June	Hisepta	September–November
(5) Sow outside late April and transplant early June	Hidena Jupiter	November–December
(6) Sow outside mid-May and transplant end June	Avon Coronet, Celtic, Celsa, Aquarius	December–February

Early summer cabbage: By sowing into pots or blocks in late February in heat, plants can be transplanted in mid-April and pointed-headed varieties such as Hispi will be ready for cutting in mid-June. Round-headed varieties such as Marner Allfruh mature about a week later, approximately seventy days after transplanting.

Summer cabbage: For cutting in late July until September, seed of varieties such as Market Topper, Stonehead and Minicole are sown under cloches or in cold frames from mid- to late March, and the plants put into the garden in late May at a close spacing of 12 × 12 in. (30 × 30 cm) Minicole especially will stand for up to three months after it is ready without serious deterioration.

Autumn cabbage: When sown in early May and transplanted in mid-June, varieties such as Hisepta will mature in September and will stand up to two months after the heads are mature. For November-December cutting varieties of the winter white type such as Hidena or Jupiter are sown outside in late April. This type can be harvested for storage in November.

Winter cabbage: For December to February cutting, varieties such as Avon Coronet, Celtic, Celsa and Aquarius which mature in succession are sown outside in mid-May and transplanted at the end of June.

When the ground is not immediately required for other crops the cabbage stumps can be cross-cut with a knife after harvesting to encourage a second crop of small heads to grow.

Calabrese (or green-sprouting broccoli)

Heads of this increasingly popular vegetable can be obtained from July until October by the use of a number of varieties and sowings. In general the best spears are produced on plants which have been sown where they will mature, but plants can be raised in a seedbed if care is taken to minimize the check to growth after transplanting.

The earliest sowings can be made in late March or early April with an early-maturing variety such as Express Corona. Further sowings can be made at 3–4 week intervals until early June using a single variety such as Express Corona, Green Comet, Green Duke or Premium Crop. All four of these varieties can also be sown together as they mature in succession from a single sowing over a period of several weeks. The terminal spear should be cut before the flower buds start to break open; the smaller lateral spears will then develop and will help to provide a succession of pickings until the first frosts.

Carrots

Carrots can be pulled from the end of May until November and, indeed, throughout the winter if the rows are protected from frost. To produce roots over this long period, a minimum of three sowings is required using different varieties. For the earliest crops a sowing of an early variety such as a selection of Amsterdam Forcing or Early Nantes is made in cold frames or under cloches in January to late February, and the largest of the young roots can be selectively pulled from mid-May until July. A sowing should then be made in the open in April of an Amsterdam type such as Colora, or one of the selections of Chantenay Red-Cored to produce roots from August until November. For late crops maturing in November and December, and for storage, an Autumn King type such as Vita Longa can be sown in May. The roots of this type are somewhat more frost-hardy than those of the Chantenay type. However, both can be stored if they are protected against frost.

Many gardeners, however, prefer to sow frequently to ensure a constant succession of young plants. To achieve this, sowings of the varieties mentioned can be made at 3–4 week intervals until mid-July. If cloches are available, a final sowing can be made in August and the plants covered in September or October in order to produce young roots in November and December. For the later sowings an early-maturing variety such as Early Nantes should be grown.

Cauliflower

Except for the mid-winter period cauliflowers can be cut throughout the year in most parts of the country, and sowing times and varieties can be conveniently grouped according to the season of production. A summarized continuity programme is given in Table 2.5.

Early summer: For cutting in June and July varieties of the Alpha and Mechelse type are sown in cold frames in the first week of October for transplanting as early as conditions will allow in March. Slightly earlier maturity will be obtained if the plants are raised in peat blocks or pots. They will be closely followed by plants raised in a seedbed and transplanted at the same time. An alternative method, if a heated glasshouse is available, is to sow the same varieties in mid-January, the plants being hardened off before planting in March or early April.

Table 2.5 Sowing and varietal sequence for cauliflowers

Method of growing	*Varieties*	*Harvest period*
(1a) Sow in cold frame early Oct. and transplant March *or* (1b) Sow in heat mid-Jan. harden off and transplant in March	Alpha, Mechelse-Classic, Dominant (in order of maturity)	mid-June to mid-July
(2) Sow under cold glass in March and transplant mid-May	Nevada Dok	mid-July to mid-August
(3) Sow in late April and transplant mid-June	Nevada Dok	late August and September
(4) Sow in mid-May and transplant early July	Flora Blanca Orco Barrier Reef	September–October October October–November
(5) Sow early May and transplant late July (for S.W. coastal regions only)	St. Agnes St. Buryan St. Keverne	late December–January February–March late March–April
(6) Sow in late May and transplant late July (regions other than S.W.)	Angers No. 2 – Westmarsh Early, Walcheren Winter-Armado April, Walcheren Winter-Markanta, Walcheren Winter-Birchington, Angers No. 5	March April late April late April–May late May

Late summer: For harvesting mid-July to mid-August a sowing should be made in March under cold glass, the plants being transplanted in the middle of May.

Early autumn: Plants from a late April or early May sowing and transplanted in mid-June will be ready for cutting in late August and September.

Late autumn: Sown in mid-May and transplanted in late June or early July, varieties of the Australian and Flora Blanca types will mature from September to November if the late autumn weather is not too severe.

Winter: Varieties of the Roscoff type can be grown for the November to April heading period *only* in mild coastal areas of the south west. They should not be grown elsewhere. One sowing is made outdoors in early May and transplanting is done in late July.

Late winter-spring: For areas other than the south-west, varieties of the Walcheren-Winter type can be sown in late May or early June and the plants transplanted in July. Different varieties will mature from March to early June (see Table 2.5).

Mini-cauliflower

Mini-cauliflower curds are high quality, complete, mature curds deliberately induced to be small (1½–3½ in. in diameter) by growing certain varieties at very close spacing. Early summer varieties of the Alpha, Danish and Snowball types have given excellent results when grown at spacings of 6 × 6 in. (15 × 15 cm) in a square arrangement (to reduce the edge effect where the plants on the outside, with more space, grow much larger). Either a single or a number of varieties will be ready for cutting in a very predictable time under this form of production and continuity can be obtained by making successional sowings. As the time taken to reach maturity varies with the seasonal trend in temperature, this factor must be taken into account if continuous production throughout the summer is planned. This is illustrated in Fig. 2.3 which can be used as a guide to predict the approximate time of cutting mini-cauliflowers using the variety Alpha sown on different dates. With this variety the earliest crops can be cut in the first week of July and continuity can be achieved with sufficient sowings until October. Plants from each sowing will mature

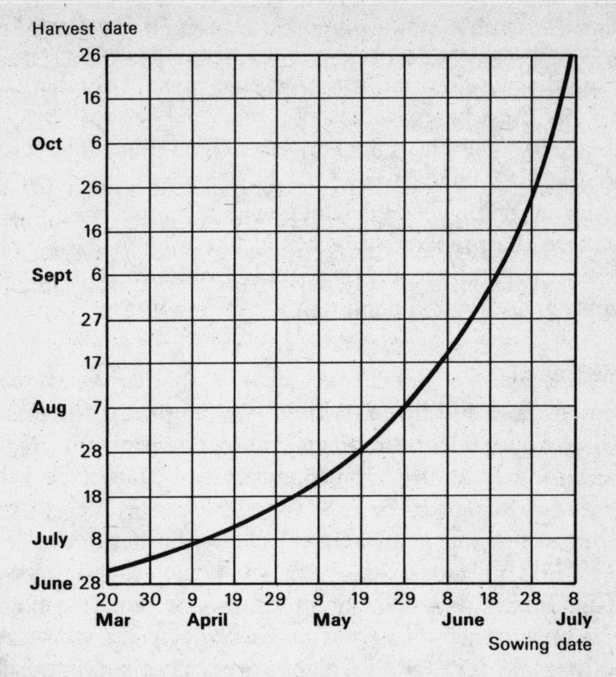

Fig. 2.3. Guide for sowing and harvesting dates for mini-cauliflower.

very uniformly, and so several sowings will be needed to maintain supplies over any length of time.

Because of their size, mini-cauliflowers are excellent for freezing whole.

Celery

Self-blanching varieties of celery such as Golden Self-Blanching, Avonpearl or Lathom Self-Blanching are grown for use from mid-July until October. The seed is sown in March or early April on the surface of compost under heated glass. After pricking-off into trays, pots or blocks the plants are hardened off and transplanted from early May to early June. If heated glass is not available, cloches can be used for raising plants which are transplanted in early June, but they will mature later than those raised in heat. Celery of the American Green type can also be grown in the same way.

To supply the period from October until February the trench types must be grown. White, pink and red varieties are available, for example, Giant White, Giant Pink and Giant Red, the latter being the most hardy. Plants are raised in the same way as for the self-blanching celery, before being transplanted into the prepared trenches in late May or early June. The celery sticks are ready for digging from October onwards and will remain in good condition for 2–3 months.

Chinese Cabbage

Until quite recently this crop could only be produced reliably in the UK for a short period from mid-September until October, because the varieties available were likely to bolt to flower if seed was sown before June. However, recent experimental work has shown that some of the newer varieties such as Tip Top and Nagaoka can be sown as early as late April and May. Sown directly into blocks or pots in a greenhouse or polytunnel where the temperature is not allowed to fall below 50°F (10°C) the plants are transplanted outside 2–3 weeks later after they have been hardened off. The earliest heads should be ready for cutting from the end of June from the late April sowing, and if sowings are made in this way at 2-week intervals until the end of May, heads will be produced until late August.

Sowings can be made outside *in situ* from mid-June until mid-August and heads will be ready from September onwards. It is advisable to sow very small areas at 10-day intervals because the plants quickly bolt after they are ready for cutting. Three suitable varieties for this season of production are Nagaoka, Sampan and Pe-Tsai which take from 9 to 12 weeks to reach maturity, depending on the weather conditions.

The heads of Chinese cabbage will keep for several weeks in a refrigerator, but they are not frost-tolerant like normal cabbage.

Leeks

This crop, which can survive even the coldest of winters, is grown mainly for use from November until March and April. However, crops can be produced as early as September if required by sowing an early variety such as Autumn Mammoth-Early Market, either in heat under glass in February, or under cloches or in a cold frame in February if weather conditions permit. These plants should be transplanted in April or early May.

Maincrop sowings are made outdoors in March, or under cold glass if soil conditions are not fit, and transplanting should be done by June. Varieties for November to March maturity include the newer ones such as Autumn Mammoth-Argenta and old favourites such as The Lyon for late autumn production. For use from Christmas to March or April, Musselburgh is still one of the hardiest of leeks, while newer varieties such as Autumn Mammoth-Herwina and Giant Winter-Catalina stand well until the late spring. For very late lifting in April and May Autumn Mammoth-Snowstar is one of several newer varieties which are suitable.

The largest transplants are normally ready for harvesting first, so if the transplants are planted in order of size along the rows the mature leeks can be lifted from one end of the row or bed, keeping the later-maturing plants in a tidy and small area.

Lettuce

It is difficult to achieve continuity of production of hearted lettuce from spring to autumn without having gluts and temporary shortages. A number of successional sowings must be made as varieties do not differ sufficiently in their cutting time to help maintain continuity. Gluts occur when two sowings are ready at the same time or when too much seed has been sown, shortages occur when sowings have not been made sufficiently frequently or when the emergence of seedlings from a sowing has been delayed because of drought or high soil temperatures (see p. 43).

To minimize these disruptive effects on continuity every effort should be made to get the seedlings of each sowing to emerge quickly and reliably, and to time the successional sowings to allow for the prevailing weather conditions. To ensure quick emergence of this crop, water the bottom of the seed drill *before* sowing, and if the weather is hot, shade the soil and sow in the early afternoon. Alternatively, the seed can be pre-germinated and fluid-sown. Plants can also be raised in peat blocks or in pots and transplanted. All of these different techniques have been described in Chapter 2 of *Know and Grow Vegetables*.

Although a lot of experimental work has been done to improve the timings of sowings for commercial lettuce production, a simple system is recommended for the garden: make sowings in succession when seedlings of the previous sowing have just emerged, take precautions to reduce high soil temperature effects, and do not sow too much seed at any one time. This method will allow for the effects of prevailing weather and soil conditions on germination and emergence of the seedlings which are major causes of gluts and shortages. Sowing at fixed time intervals such as 7–10 days does not allow for the vagaries of the weather and causes the problems of over- and under-supply.

The earliest crops are produced in late May and early June from plants raised under glass and transplanted at the end of March or early April under cloches or into the open ground. Continuity is then maintained throughout the summer and early autumn by sowing at intervals from March to July as described above. Among varieties suitable for successional sowing over this period are Avondefiance and Hilde II (butterhead type), Little Gem and Lobjoits (cos), and Webbs' Wonderful and Great Lakes (crisphead). A general indication of the harvest dates of butterhead varieties from different sowing dates is given in Fig. 2.4. In sheltered positions or when cloches are available, sowings are made in late August of varieties such as Valdor or Winter Density which will be ready in late May. Plants from a further sowing made under cloches

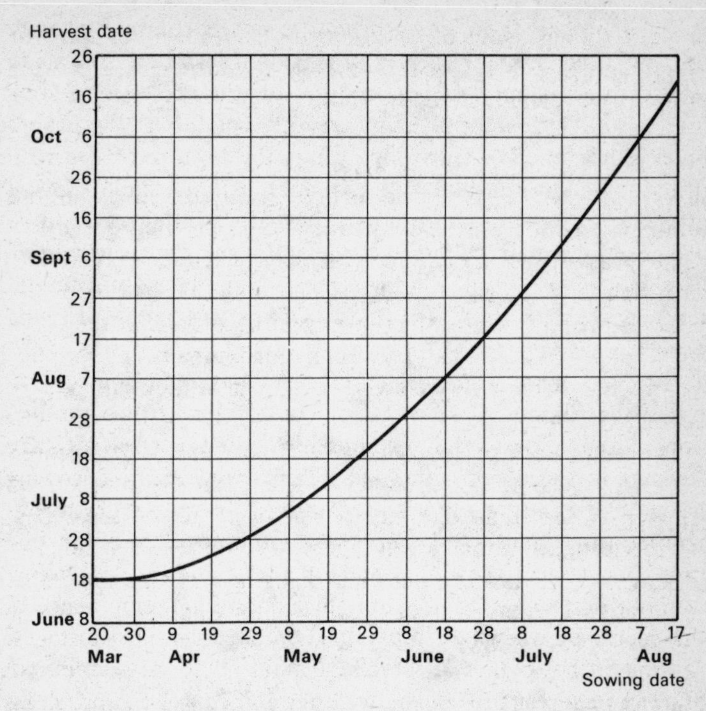

Fig. 2.4. Guide for sowing and harvesting dates for hearted butterhead lettuce. Crisphead varieties will be up to three weeks later maturing.

Table 2.6 Sowing and varietal sequence for outdoor hearted lettuce

Method of growing	*Varieties*	*Harvest period*
(1) Sow in cold frames in blocks in late Feb. and transplant in early April under cloches and open ground	Avondefiance Hilde II Little Gem Great Lakes	late May–June
(2) Sow at intervals (see text) from March to July *in situ* or into blocks and transplant	Avondefiance Little Gem Great Lakes	June–October
(3) Sow under cloches or frames in Oct. and transplant Feb. or March under cloches	Premier	early May
(4) Sow outdoors in mild or sheltered areas in late August	Valdor Winter Density	late May

or cold frames in October are transplanted in February or March under cold glass for early May cutting. The sowing programme is summarized in Table 2.6.

Leaf lettuce

If lettuce *hearts* are not necessarily wanted throughout the summer, growing unhearted plants should be tried because it is very easy to get a continuous supply of lettuce leaves. The term 'leaf lettuce' is generally used to describe lettuce of the Salad Bowl type where the leaves can be picked as required, but a supply of leaves can be obtained from many cos varieties by growing them at very close spacing so that they cannot produce hearts.

The equivalent of four or five normal-hearted lettuce per week from May to mid-October can be obtained from sowing an area of less than one square yard on each of ten dates. After cutting, the stumps will regrow to produce a second crop of leaves. Full details of sowing dates and of this system of growing lettuce leaves have been given in Chapter 1 of *Know and Grow Vegetables*.

Marrows and courgettes

Marrow plants produce a succession of fruit provided they are picked regularly, and only two sowings are needed to supply courgettes or marrows from mid-June until October. The earliest crops are obtained from sowing in early April into pots or blocks under cold glass or on a windowsill, the plants being transplanted under cloches in late April or early May after hardening off. A second sowing is made from mid- to late April for planting in early June. The bush types produce fruit 2–3 weeks earlier than the trailing type and should be used for early crops. For courgette production, where the fruit are picked small and immature, varieties such as Green Bush F_1 and Zucchini are suitable.

Onions

With the recent introduction of the early-maturing, intermediate day-length varieties (see Chapter 6) it is now possible for the gardener to produce bulb onions from early June to October and to have them available throughout the year with storage. The crop can be produced from seed sown directly in the garden, from transplants, or from sets (see Table 2.7).

The earliest crops to mature from late May onwards are grown over winter from sowings made in August. For this form of *early* production only the varieties of Japanese origin must be grown, and to obtain a succession of mature onions the following are suggested: Express Yellow O-X (harvest in early June), Imai Early Yellow (mid-late June) and Senshyu Semi-Globe Yellow (late June–July). The sowing date in August is of critical importance, for if sown too early for the district, the plants will grow too large before the winter and may bolt in the spring; if sown too late, the plants may not survive the winter or will only produce small bulbs. In Scotland sowing should be done in the first week of August; in the north the second week; in the Midlands and East Anglia the third week; and in the south the fourth week of August. The aim should be to get the seedlings above ground within 7–10 days, so water the drill before sowing if the soil is dry.

The introduction of this early form of production has reduced the need in most areas to raise transplants from sowing in heat in January, and after hardening off, transplanting in late March or early April. This transplanting method should still be used when growing bulbs for exhibition purposes. It can also be used when the soil is heavy or poorly drained, for under these conditions plants will not overwinter satisfactorily. Recently, experimental work has shown that up to five plants can be raised in a 1¾ in. square (4.3 cm) peat block. They are not separated but are planted out all together. The advantages of this system are a saving in block costs, much higher yields and about 2 weeks earlier maturity than crops sown directly outside.

Mature bulbs can be produced in August from planting onion sets of varieties such as Stuttgart Giant or Sturon in late March or early April. If small-sized or heat-treated sets

Table 2.7 Sowing and varietal sequence for bulb onion continuity

Method of growing	*Varieties*	*Harvest period*
(1) Sow outside in August	Express Yellow O-X Imai Early Yellow Senshyu Semi-Globe Yellow	early June mid-late June late June–July
(2) Sow in heat in Jan., harden off and plant April	Rijnsburger-Wijbo	August
(3) Plant sets in late March or April	Stuttgart Giant Sturon	August
(4) Sow outside in March or early April	Hygro Rijnsburger-Balstora	September–October

(see Chapter 6) are used, this method is very reliable and convenient but the temptation to plant the sets in February or early March should be resisted for the plants may bolt.

For harvesting in August, September and October, the main crop should be sown outside in March or early April as soon as soil conditions are fit. Out of many recently-introduced varieties Rijnsburger-Adina, Rijnsburger-Wijbo, Hygro and Rijnsburger-Balstora mature in that order; the last two varieties are recommended for their storage qualities.

Onions spring or salad

Special varieties of onions such as White Lisbon and White Lisbon-Winter-Hardy are grown closely together, and with successional sowing can be pulled from April until the autumn. The earliest crops in April are grown from sowings made in the open ground from mid-August to early September, and pulling will be possible from these sowings until June. The varieties are very hardy and will survive most winters provided they are grown on well-drained soil. The crop can be advanced by protecting with cloches for all, or the latter part, of the winter.

For a supply of salad onions from June onwards the same varieties should be sown under cloches in late February or early March followed by successional sowings in the open ground at fortnightly intervals until June. The timing of these sowings is not critical because with close spacing the plants when ready will hold for several weeks.

If overwintered bulb onions are sown thicker than required as an insurance against winter losses, thinnings from this crop can also be used for salad purposes from January to April.

Parsley

Parsley can be picked throughout the year if two sowings are made, one in March and a second in June or July for early spring use. Plants from this latter crop can be covered with cloches from December onwards to encourage earlier leaf growth in the spring. Because the seeds can take many weeks to germinate in the soil there are considerable advantages from pre-germinating the seeds indoors. They are then either fluid-sown directly into the garden soil, which can save several weeks, or pricked off into blocks and subsequently transplanted. Recommended varieties are Bravour, Moss Curled and Paramount-Imperial Curled.

Parsnip

Because this crop is very hardy and the roots store well in the soil without deterioration over the winter period, a single sowing made any time between late February and May will provide roots from October until March. Very early crops of medium-sized roots can be produced in August or September by pre-germinating the seed and fluid-sowing as soon as soil conditions permit. For early production the plants should be widely spaced (2 plants per square foot) so that they do not compete with each other and fast growth should be encouraged by every possible means. Suitable varieties are Avonresister and White Gem.

Peas

With a combination of sowings and different varieties, peas can be picked from the end of May until late autumn. The earliest crops are produced from sowings made in October or November on well-drained soils in mild areas or, in the north, with cloche protection from December. Round-seeded, hardy varieties are used such as Feltham First or Meteor which mature from late May onwards. To follow, the same round-seeded varieties can be sown under cloches in February if soil and weather conditions permit. It is a waste of time, seed and effort to sow these early varieties under cold, wet conditions for they will not germinate, but only rot even when the seed has been dressed with fungicide.

The sweeter, wrinkle-seeded varieties can be sown from late March until late May giving a succession of crops until August. If four varieties which differ in their time of maturity are sown together, continuity is reliably obtained over a short period and fewer sowing dates are needed. For example, if short lengths of row of the following four dwarf maincrop varieties are sown on the same date their combined picking periods would last 2–3 weeks because they mature in succession: Hurst Beagle, Hurst Canice, Victory Freezer, and Hurst Greenshaft. Therefore a sowing of these varieties every three weeks will provide a succession of crops until August. These newer, dwarf varieties tend to produce most of their pods together on the plants and so do not provide the same long period of picking often found with the older, taller, indeterminate varieties. The timing of the sowings is therefore more

Table 2.8 Sowing and varietal sequence for peas

Method of growing	*Varieties*	*Harvest period*
(1) Sow outside in mild areas or under cloches in Oct. or Nov.	Feltham First Meteor	late May–June
(2) As (1) in February or early March	Feltham First Meteor	early to late June
(3) Sow outside in succession from March to late May (see text)	Hurst Beagle Hurst Canice Victory Freezer Hurst Green Schaft	late June–August
(4) Sow outside in June	Kelvedon Wonder	September

important and the keeping of detailed records of sowing and picking dates and daily temperatures will be of great help in planning a sowing programme for your site in future years.

In June or July depending on the district, sowings can be made of a variety such as Kelvedon Wonder for picking in September, and even into October with the use of cloches.

Potatoes

Potatoes can be lifted from late May until late autumn by the traditional practice of growing both early and maincrop varieties. In mild areas, or with some form of protection such as cloches or low polythene tunnels, 'new' potatoes can be lifted in late May, and if a warm greenhouse is available earlies can be forced to provide new potatoes in April or even earlier.

'Chitting' or sprouting of the seed potatoes before planting is essential for early varieties and will enable lifting to be 2–3 weeks earlier, while chitting of maincrop varieties will result in heavier yields if lifted in late summer (see Chapter 6, p. 191).

For the earliest crops the chitted seed tubers are planted in mid- to late March, the shoots being protected from frosts during the early stages of growth by earthing-up. All varieties should only be planted when soil conditions are fit and when the soil temperature at a depth of 4 in. (10 cm) exceeds 43°F (6°C) on more than three consecutive days. For a speculative crop to harvest in December, tubers lifted from an early variety can be replanted in late July and covered with cloches or a tunnel in September. Varietal and planting date suggestions for traditional methods of providing a continuous succession of potatoes are summarized in Table 2.9.

However, recent research has suggested an alternative way of producing both early and late crops using a single variety. These can supply potatoes from late May until the autumn. This is achieved by manipulating the 'physiological age' of the seed tubers, as reflected in sprout development, by means of the temperature regime they experience before planting. The concept of physiological age is described in Chapter 6 (p. 191).

Table 2.9 Planting and varietal sequence for potatoes

Method of growing	*Varieties*	*Harvest period*
(1) Plant early varieties outside in mild areas or under cloches mid-March	Maris Bard Ulster Sceptre Pentland Javelin	late May–June
(2) Plant early varieties early April	as (1)	late June–early July
(3) Plant second early varieties mid-April	Maris Peer Red Craigs Royal	July–August
(4) Plant maincrop varieties late April–early May	Desiree King Edward Maris Piper Pentland Crown	September–October
(5) Plant an early variety late July and protect in September	Arran Comet Ulster Chieftain	December

An example of the method with the early maincrop variety Desiree is as follows:-

In order to produce 'physiologically-old' seed tubers at planting it is important to obtain seed tubers in late autumn or early winter before they break dormancy. The tubers should be stored apical-end uppermost in daylight at a temperature of about 50°F (10°C) until just before planting, when the temperature may be reduced to 'harden' the tubers and minimize any temperature shock at planting. This will give an early crop. In contrast, a late-maturing crop can be grown from 'physiologically-young' tubers which are produced by keeping tubers cool, less than 40°F (4°C) but above freezing, until planting.

Radish

Radishes are one of the quickest and easiest crops to grow and to maintain a supply from May to October. The seed germinates well and it is all too easy to sow too much each time; a short length of row at fortnightly intervals will give good continuity. The earliest sowings of forcing varieties such as Saxa and Saxerre are made in cold frames or under cloches in February and March to pull in May. In the open, successional sowings of varieties such as Cherry Belle and French Breakfast can be made from March until September. Another method is to sow seed of mixed varieties which mature at different times; fewer sowings are then needed. The plants should be pulled young, otherwise the roots can become tough and hot to the taste. This is another reason why small quantities of seed should be sown at frequent intervals. Late sowings can again be protected by cloches.

The large-rooted, winter radish should be sown in July or early August and are thinned to 8 in. (20 cm) apart. They are quite hardy and can be left in the ground for lifting as required throughout the winter although as a precaution they should be covered with straw during severe weather. Alternatively, the roots can be lifted and stored in November. Suitable varieties available are China Rose, Black Spanish Round and Mino Early.

Spinach

This crop can be obtained virtually throughout the year. In a similar way to radish, spinach needs to be sown little and often from March until July to provide a succession of young leaves from May until October. The earliest sowings are made under cloches at the end of February if the soil has warmed up, and in the open from March onwards at 2–3 week intervals until July using varieties such as Longstanding Round and Sigmaleaf. The largest fully-grown leaves are taken frequently so that they do not get tough.

For winter production from October until May, winter spinach varieties such as Greenmarket or Sigmaleaf are sown on two or three occasions in August and September on a well-drained site. Best results are obtained if the plants are protected by cloches from October throughout the winter period.

Swede

A single sowing of this crop in May will provide roots from late August until March direct from the ground, for the plants are extremely hardy. Because of possible problems with mildew the sowings are made in early May in the north of the country and late May or even early June in the Midlands and the south. A suitable variety is Marian.

Sweet corn

For best results this crop requires a warm, sunny, sheltered site because it does not start to grow until the temperature rises above 50°F (10°C) and it is also susceptible to frost damage. Except in the southern part of the country where the seed may be sown directly into the soil in mid-May, it is best to raise young plants in blocks or pots and transplant after the risk of frost has passed.

Seed can be pre-germinated from mid-April onwards and the chitted seed carefully transferred to pots or blocks of compost. Three to four weeks are then needed at a temperature of 50 to 55°F (10–13°C) to grow the plant to a size ready for transplanting. Early crops can be planted under the protection of cloches or low tunnels in mid-May if the soil has been previously covered. These can be followed by plantings in the open ground in June after the risk of frost is over, or be prepared to protect from frost. The use of clear polythene mulch will advance the maturity of this crop by up to a month. The earliest crops of varieties such as Kelvedon Sweetheart are ready for picking in late July or August depending on the season. Later varieties such as Aztec and Northern Belle will mature in September. The cobs should be picked when they are 'milky', before they become over-mature.

Tomatoes, outdoors

Tomatoes, like Sweet corn, are easily killed by frost and need a warm, sunny position if they are to do well outside. The newer bush or dwarf varieties are more tolerant of cold conditions than the older, staked varieties and have become more popular for they can be grown successfully even in Scotland.

The normal method of growing the staked plants out of doors is to sow the seeds under glass or on a windowsill in March or early April in a temperature of 60–65°F (16–

18°C). The seedlings are pricked off into blocks or pots, and after hardening off are planted outside in early June or when the risk of frost has passed. If cloches or tunnels are available the plants may be transplanted in mid-May provided the soil has been covered and warmed up beforehand. When the plants reach the glass or polythene they are uncovered and staked. A suitable variety is Alicante, which will mature from late July onwards.

Plants raised in this way can also be transplanted into large containers or growbags and placed on paths, terraces or patios, or against a south-facing wall. With these staked varieties the plants should be 'stopped' after four trusses, or in early August. Picking can be continued until the first frosts. The remaining green fruits should be brought inside by mid-September for ripening.

Plants of the bush varieties may be raised and planted in a similar way to that described but are not normally staked or pruned of their side-shoots. An alternative method is to sow pre-germinated seeds thinly in a single row under tunnels or cloches in mid-April. The seedlings should emerge 7–14 days later and if sown thinly no thinning of the plants will be required. The protection can be removed in June. In southern England pre-germinated seeds can be fluid-sown directly into the open ground in late April sowing two or three seeds every 12 in. (30 cm) along the rows. With both methods using pre-germinated seed there is a risk of damage from late spring frosts, but experience has shown that the risk is low in southern counties. Even if frosts do occur, growth is not severely impaired provided the night temperature does not fall below −2°C. Suitable bush or dwarf varieties are Sigmabush, Sleaford Abundance and Alfresco and, depending on the season, ripe fruit can be picked from mid-July onwards. Many believe that the flavour of these bush varieties is better than that of the staked varieties.

Turnips

Roots of this crop can be pulled for most of the year from a

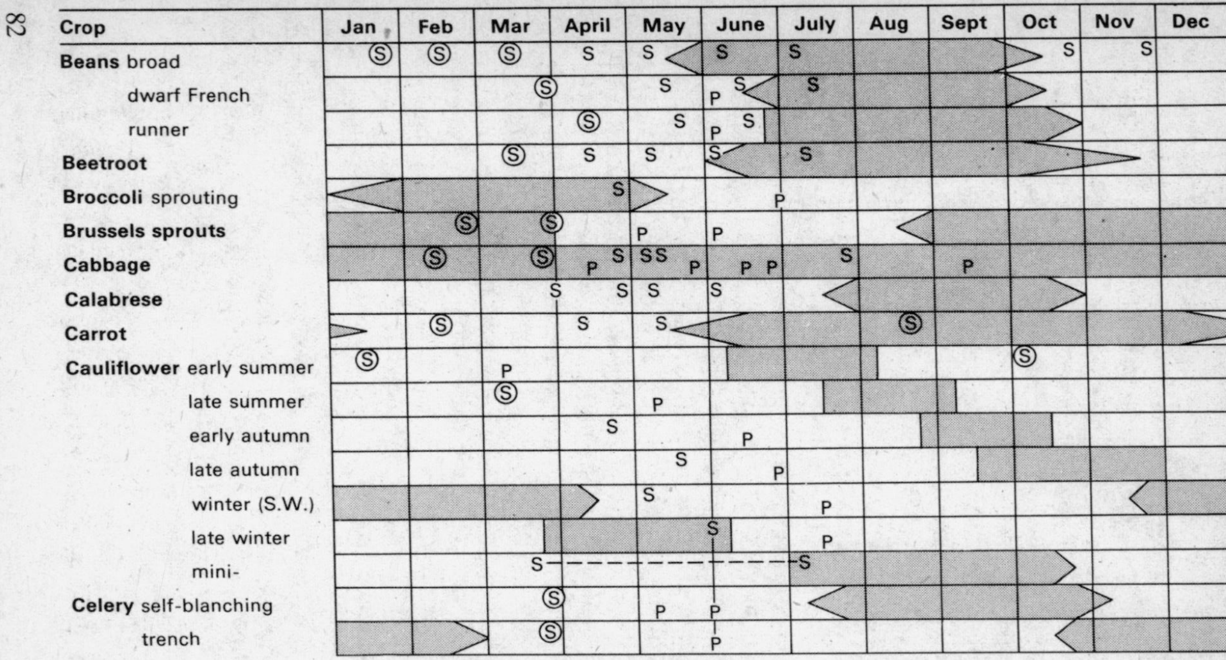
Crop
Jan
Feb
Mar
April
May
June
July
Aug
Sept
Oct
Nov
Dec
Beans broad
dwarf French
runner
Beetroot
Broccoli sprouting
Brussels sprouts
Cabbage
Calabrese
Carrot
Cauliflower early summer
late summer
early autumn
late autumn
winter (S.W.)
late winter
mini-
Celery self-blanching
trench

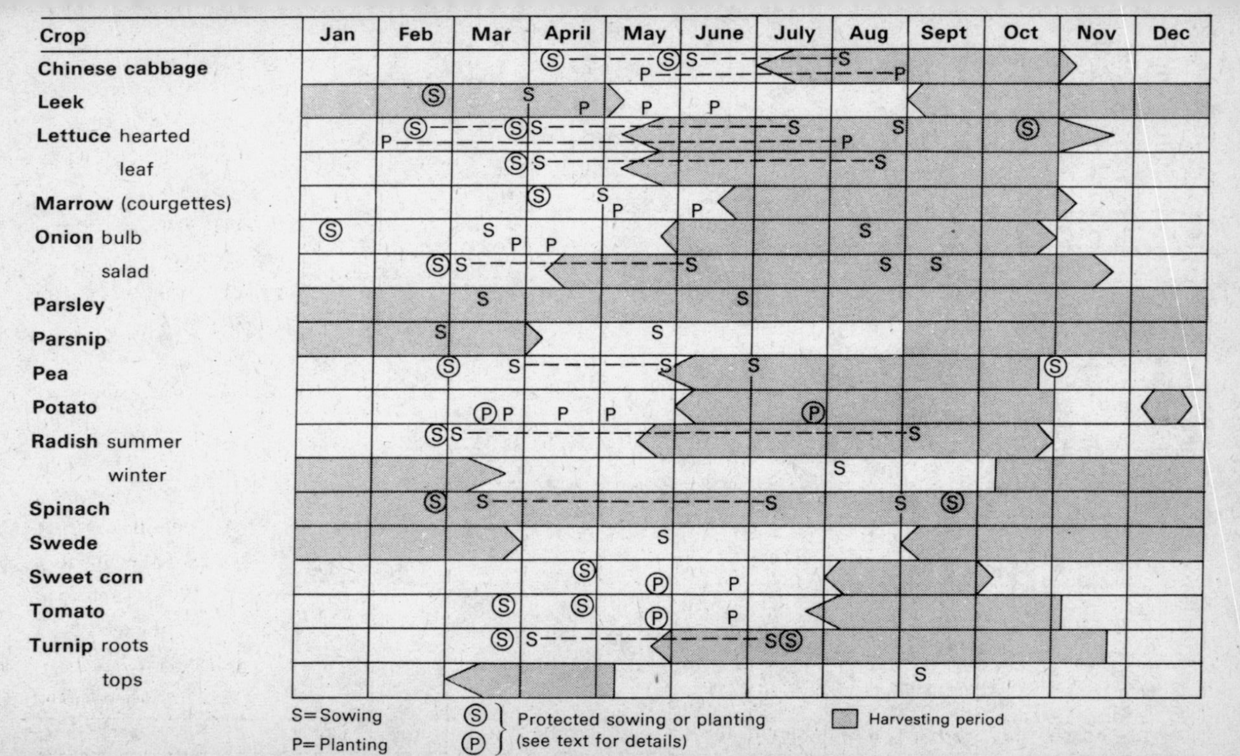

Fig. 2.5. A summary guide to plan crop continuity.

succession of sowings using varieties suited to the season. The earliest crops are produced in late May or early June from a sowing made under cloches in March of varieties such as Purple Top Milan, Snowball and Sprinter. Successional sowings of the same varieties can be made at 3–4 week intervals until July for pulling in sequence until October.

To provide roots for late autumn and winter, seed of maincrop varieties such as Golden Ball are sown in July and the roots are best lifted and stored in November for use throughout the winter.

The tops of turnips are also grown for cutting as 'spring greens'. These crops are sown in August or September and the tops harvested in March and April but if these plants are protected during the winter earlier pickings of 'tops' can be made.

3 Storing vegetables

There are various methods of storing vegetables which can extend your season of supply. These range from the simple, such as putting lettuce heads into plastic bags before placing them in a refrigerator or other cool place to keep for a few hours or days, to the equally straightforward but rather more time-consuming methods used to prepare crops such as onions and beetroot for up to 6–8 months storage. Furthermore, the availability of certain crops can be extended even more if they are preserved by bottling, canning, freezing, pickling and salting, but these subjects are beyond the scope of this book.

Vegetables such as potatoes and onions are easy to store; leafy vegetables on the other hand are more difficult, but the aim remains the same, to keep them as nearly as possible in the state they were at harvest so that they can be enjoyed 'fresh'.

GUIDELINES

Once harvested, vegetables deteriorate, some very rapidly as they are removed from their natural supply of food and water. This means they should be harvested only when there is some good reason for not leaving them in the ground. Fast-growing lettuce and cauliflower, for example, become quickly over-mature and eventually bolt if left uncut and temporary storage is a way of extending the supply of these crops. Similarly, tender vegetables are best harvested and stored before low night temperatures cause damage and wastage through 'chilling injury' (see p. 105). However, many root crops, leeks, and most brassicas, can be left in the ground until required as they can stand all but the heaviest frosts. Providing a little frost protection can be less trouble and less wasteful than harvesting and storing these crops. So the first guideline to follow is:

> Harvest and store crops only when they cannot be left in the ground.

Vegetables deteriorate gradually after harvest, using up their in-built supplies of food and water until they become inedible. Because the rate of deterioration is slower at low temperature, many vegetables can be successfully stored in a refrigerator. But each vegetable has its own particular storage requirements and whereas conditions in a refrigerator may suit leafy vegetables, it is not the best place to store tomatoes or potatoes. For this reason the storage requirements of the main groups of vegetables will be discussed separately. In all groups of vegetables it is obviously only possible to get good quality vegetables out of store if they are of good quality when they are put in. In this context good quality means well-formed and undamaged. It follows that three further guidelines should be:

> harvest carefully and selectively
> cool rapidly
> store in suitable conditions

STORE NATURALLY – IN THE GROUND

Many brassicas and root crops are ready for harvest by early winter, but the outside temperatures from November to March are generally low enough to restrict further growth and development. At these outdoor temperatures these vegetables do not deteriorate very quickly and they are best 'stored' where they grow, in the ground. This works well for the many vegetables which are reasonably hardy and there are the advantages that we use free, natural refrigeration and also avoid the wastage which commonly develops on tissues damaged at harvest.

Successful storage in the ground also depends on good control of pests and diseases. Bad attacks of aphid, carrot fly, and other pests can reduce the quality as well as the yield of vegetables drastically, and where these occur it may be best to harvest, grade and store or deep-freeze the pest-free surplus supplies.

The winter brassicas which are suited to leaving in the ground (see Chapter 2) are not normally given additional protection from the weather, though they may lose some outer layers of leaves because of freezing injury. Vegetables growing at, and below ground level, such as carrots, beetroot, swede, turnip and leeks can be given some additional protection from the weather by simply drawing up the earth around them. This and other methods are described in detail later.

HARVEST CAREFULLY

Vegetables are made of delicate tissues that are easily damaged when handled at harvest time. Broken tissues exuding sap are an ideal medium for the growth of disease micro-organisms and it is not surprising that the greater the level of damage at harvest the greater the level of wastage that occurs in store. Harvest damage not only wastes edible tissues, it greatly increases the chance of fungal and bacterial infection which can affect the whole vegetable and often adjacent ones in store. Any vegetable that is badly damaged during harvesting is not suitable for storage and should be used up first.

When vegetables are cut or trimmed during harvest the knife used should be cleaned from time to time, as this will reduce the possibility of infection of the wound. The knife should also be sharp enough to make a clean cut, as these wounds dry out and callous over more quickly. Tearing and bruising tissues at harvest should be avoided, as such wounds do not dry out readily. Dropping and banging vegetables together at harvest should be avoided because this causes the breakdown of tissues just under the surface and leads to bruising. Such damage is especially difficult for the plant to repair if the surface is broken and the sap is exposed to infection. The exception is the leaves of root crops which can be removed by twisting off at harvest.

How delicate are vegetable tissues and what happens when they are roughly handled? Three examples are described to illustrate the increasing susceptibility to damage.

Onion

The onion bulb is a natural storage organ and by the time it is fully grown and dried the edible, fleshy scales of the bulb are surrounded by two or three layers of protective, dried, papery skins. The bulb is then an ideal storage vegetable – a package of dormant, edible tissues, ready-wrapped for storage, and it will remain in a state of rest for many months if properly treated. The length of the storage life can easily be reduced, however, by rough handling. The papery skins protect the bulb from damage and also prevent the fleshy scales from drying out. Even the partial loss, say 20 per cent of the area, of papery skins to reveal the cream-coloured fleshy scale underneath will lead to softening and shrivelling of the bulb after a few months in store. Severe bruising will lead to premature sprouting, but even dropping a bulb as little as 8 in. (20 cm) onto a hard surface produces a bruise which extends several scales deep into the bulb.

Carrot

The fleshy tap-root is another storage organ, but one that is not easily harvested without damage. The surface of the root is covered by a very thin skin of dead cells, about half as thick as a human hair. It is specially produced to protect the surface from abrasion by soil particles during growth, but it is also a useful protective barrier during handling. Its usefulness is limited, however, as it is easily scraped off, and during lifting and washing about 20 per cent of the skin of the carrot may be removed. The skin can and does regenerate, but this takes time and is usually accompanied by browning of the surface and the development of bitter off-flavours. It is for these and other reasons that storage in the ground is preferred for this crop.

In addition to skin damage, normal lifting by hand produces serious cuts and abrasions in a proportion of the roots. This type of damage is difficult to repair and frequently never is, as micro-organisms grow readily on the broken tissues. Thus it pays to put on one side carrots damaged in this way for immediate use. Carrot leaves are usually cut short at harvest and the tip of the taproot is also broken. These damaged surfaces also occasionally provide access for rotting organisms.

The choice of cultivar and various cultural practices can also influence the likely damage level at harvest. Cultivars of the Chantenay and Autumn King groups generally have a better developed skin with more layers of cells than cultivars of the Amsterdam Forcing and Nantes groups and therefore stand up to handling better. Cultivars of the latter groups are regarded as having better quality because of their crisp and brittle texture, but this makes them more prone to breakage during handling and for this reason also, they are less suitable for storage. Fully-mature carrots are less prone to breakage at lifting than immature roots. Later-maturing carrots for storage or overwintering are drilled from mid-May to mid-June in central England, and are mature in most seasons from mid-October onwards.

Cauliflower

The extraordinary nature of the cauliflower curd can be appreciated not only on a plate, but also in its fresh state, with a magnifying lens. It is then seen to be extremely delicate, being composed of thousands of unprotected growing points arranged in spiral formations. Unfortunately, these growing points are easily broken by merely touching the curd, and the sap so released is an ideal medium for the growth of micro-organisms. There is no known natural mechanism to repair damage of the curd. Therefore for successful short-term storage avoid anything touching the curd, even water (heavy rain can ruin curds) and trim the outer leaves in such a way that they can be folded over to protect the curd. The cut stem surface is unlikely to rot whilst being stored provided the cut is clean and able to dry out. Cauliflowers cannot be successfully stored for more than two to three weeks because of the peculiar nature of the curd and the difficulty of handling it without damage and preventing fungal and bacterial invasion of the tissues.

The above examples show a typical range of susceptibilities to damage shown by vegetables. Particular care should be taken when harvesting vegetables that are not natural storage organs as these have the least well-developed systems for repairing damage. The benefits of storing vegetables are greater if wastage levels are low, and good storage begins with careful harvesting and the selection of good produce.

COOL RAPIDLY AFTER HARVEST

Because vegetables deteriorate rapidly at high temperatures the sooner they can be moved into cool conditions after harvest, the better. As a general guide, the rate of deterioration of plant tissue will be halved for each 18°F (10°C) fall in temperature, and so vegetables will respire less and will store much longer if their temperature at harvest can be reduced. Those vegetables which are growing rapidly at harvest in high temperatures in the summer, are likely to benefit most by cooling, providing it is done soon after harvesting.

The benefits of cooling differ from vegetable to vegetable. For example, the rapid growth of lettuce and cauliflower can be effectively stopped by cooling. Similarly, the undesirable conversion of sugars to starch in peas and sweet corn can be greatly reduced by lowering the temperature. Even vegetables which deteriorate slowly over a period of months such as turnips, swede and beetroot benefit from low temperature storage as this slows down the rate of cellulose thickening which makes them tough to eat.

The expected storage life of vegetables can be calculated by knowing the maximum storage life under ideal conditions (see Tables 3.1, 3.2 and 3.3) and the storage temperature actually available. This effect of temperature can be demonstrated by considering the storage life of a summer lettuce. It can be stored for up to 14 days in a cold refrigerator, so in round terms it uses 1 of its 14 units of food reserves a day at 32°F (0°C). As domestic refrigerators commonly run at about 40°F (4°C) it will use 2 units a day and so it can be kept for 7 days. At 50°F (10°C) it will use 3 units a day, so will store for up to 5 days in a cold cupboard or larder at this temperature. At 68°F (20°C) it will use 9 units a day, so will last less than 2 days at room temperature. These figures summarize in a simplified form the results of actual experiments, and the principles of the system work well for most vegetables.

We should only reduce the temperature of vegetables to that of the freezing-point of water 32°F (0°C). At temperatures below this, ice crystals can cause irreversible damage to plant tissues, killing the cells. Most vegetables can be cooled to this temperature safely for the heat generated by the plant's respiration prevents it from freezing. It is the ideal temperature to store brassicas, lettuce and other leafy vegetables, and for most root crops, although gardeners often have to compromise and use warmer temperatures that happen to be available. However, vegetables of sub-tropical origin such as potatoes, tomatoes, cucumber, most beans and sweet peppers show adverse effects of temperatures well above 32°F (0°C), known as 'chilling injury' and we will return to this topic later in the chapter. Deep freezing at +5 to −5°F (−15 to −20°C) is a different way of preserving, rather than of storing, vegetables. This temperature level not only kills all the cells but is low enough to suspend virtually all deterioration. This ensures that vitamins, proteins and other unstable food substances are preserved. Micro-organisms that can survive are inactive at this low temperature.

Domestic refrigerators normally run at 35 to 40°F (2 to 4°C) and hence there is little danger of freezing vegetables. A very useful storage life can be obtained for many vegetables under such conditions provided they are cooled rapidly after harvest, and it can be worth getting a second-hand refrigerator for this purpose. A bulk of vegetables is best cooled by leaving some gaps between them for the heat to escape, and cooling is therefore most effective if the produce is unwrapped. However, the atmosphere in a refrigerator is fairly dry because moisture present in the air freezes-out on the cooling coil. As a result vegetables and especially leafy vegetables wilt rapidly in a refrigerator. This is avoided by wrapping them in polythene once they are cool.

Whenever possible it is preferable to harvest vegetables in the cooler part of the day (usually very early morning) and once harvested, they should be placed in a cool place such as a cellar or in the shade of a tree. If refrigeration is not available vegetables can be cooled several degrees by the evaporative cooling of water. Water removes heat from its surroundings when it evaporates, so leafy vegetables with a large surface area such as lettuce and cabbage can be kept cool quite effectively if sprayed with water once or twice a day. This will not work if the water cannot evaporate, as when vegetables are placed in a closed polythene bag. Cooling in closed polythene bags will give rise to condensation in the bags which will mark and spoil cauliflower curds and delicate leaf tissue over a period of days.

STORAGE CONDITIONS FOR THE MAIN GROUPS OF VEGETABLES

Not all vegetables want the same storage conditions, and both the temperature and the humidity of the air in which they are stored should, ideally, be controlled if the best results are to be obtained. The importance of **temperature** has already been described.

Humidity

The water content of most vegetables is very high (90 per cent in carrots and 95 per cent in lettuce) and, until harvesting, is maintained by water being continually absorbed from the soil. Once harvested, however, the plant is removed from its natural water supply, and begins to lose water in the form of vapour to the surrounding air. If this is allowed to go on for any length of time vegetables will wilt and they only have to lose as little as 5 per cent of their weight, to become limp and unattractive. The rate of water loss from any given vegetable depends to a large extent on the water-vapour content, that is the humidity, of the air surrounding it. Of course vegetables with lots of thin leaves, like lettuce, lose water far more quickly than compact vegetables, like swedes. But just like washing on a clothes line, vegetables will dry out quicker when the humidity is low – on a good drying day – than when it is high. As vegetables have a high water content, most of them store longer in conditions of high humidity.

Even when the optimum humidity for storing a particular vegetable is known, actually attaining it, and maintaining it, is difficult because air holds much more water vapour at high than at low temperatures. This means that every time air warms up its humidity falls, and *vice versa*. Humidity is often quoted as per cent relative humidity, which expresses the water-vapour content as a percentage relative to the maximum possible, at that temperature. When air of high humidity is cooled, its humidity rises further until the point is reached when there is too much water vapour for the air to hold. At this point the air is said to be saturated with water vapour (or is at its dewpoint) and its relative humidity is 100 per cent. If cooling continues, then water vapour turns to liquid water, and this is frequently seen on the cooling surface as condensation. We see examples of this in everyday life as condensation inside windows of houses in cold weather, and as dew on the ground. It can also be seen inside polythene bags of vegetables cooling in a refrigerator.

The reverse happens when the temperature rises; for example air saturated at a low temperature (say 40°F (4°C)) will only have sufficient water to achieve 50 per cent relative humidity at 50°F (10°C), and only 10 per cent at room temperature. Such changes have a dramatic effect on harvested vegetables which are brought into a house temperature of, for example, 65°F (18°C) from an outside temperature of 50°F (10°C); under such circumstances they are liable to wilt rapidly. This example also shows that temperature and humidity must be considered together and if one wishes to store vegetables at high humidity, the simplest way to do this is to keep the temperature low and make sure that there is some free water in the container with the vegetables.

An unheated shed is a good place to store vegetables during the cooler months but vegetables will also wilt there unless they are stored at a suitable humidity. The temperature and humidity requirements for the main groups of vegetables are given below. At the beginning of each section, and in the tables, the ideal storage conditions to obtain the maximum storage life are described. These are given so that readers can see what conditions to aim for, and what storage life can then be achieved.

It is realized that most gardeners will have to compromise, especially as far as temperature is concerned, because they must use whatever facilities are available. Humidity requirements are only given as approximate guides since it is clear that humidity will vary greatly as the temperature changes, and furthermore, few people have instruments to measure it. Thus the highest possible humidity, often quoted as 95 per cent plus, is indicated as 'highest possible' in the tables, and is usually achieved by placing in polythene bags, or by using a mist spray. Humidities around the 90 per cent level are indicated as 'high', and can be obtained by leaving bags and containers open to allow some circulation of air, or by using porous paper sacks or boxes of sand in which to store the vegetables. The lower level of humidity required for onion bulbs is obtained by allowing a very free circulation of dry air.

Leafy vegetables

Leafy crops such as brassicas, celery, lettuce, spinach and

watercress should be stored ideally at 32°F (0°C) at the highest obtainable humidity. If possible leave crops such as Brussels sprouts, cabbage, celery and leeks in the ground. Other leafy crops such as lettuce and cauliflower, however, will have to be cut otherwise they will go rapidly over-mature and quality will be lost.

Vegetables in this group are not natural storage organs and with the exception of slow-growing cultivars of cabbage, generally have very limited food reserves. In addition, many in this group are growing quickly at the time of harvest, possibly at high summer temperatures and for all these reasons it is especially important to cool them rapidly after harvest to

Table 3.1 Storage conditions for leafy vegetables

	Ideal storage conditions			
	Temperature and humidity	*Freezing point °F*	*Likely maximum storage life (days)*	*Number of days of likely storage life at low ambient temperatures 40–50° F (4–10° C) in winter*
Asparagus	All of this	31° (–0.5°C)	14	
Broccoli, sprouting	group	31°	10–14	5
Brussels sprouts	require	30° (–1°C)	14	7
Cabbage, spring	32°F (0°C)	30°	7	
Dutch white	and	30°	120	60
Chinese	highest	30°	60	
Calabrese	possible		3	
Cauliflower	humidity	30°	20	4
Celery		31°	60	15
Kale		31°	14–20	7
Leek		30°	60	15
Lettuce		31°	14	
Onion, salad		30°	14–20	5
Parsley		30°	30	7
Peas (in pod)		30°	14	
Spinach		31°	14	
Watercress		31°	4	1

reduce the rate of deterioration. Once these vegetables are cooled the water content is best conserved by wrapping in polythene.

The attractive green colour of chlorophyll in the green tissues gradually breaks down in store, and turns to yellow, and this shows that the tissues are coming to the end of their useful life. The old, outer leaves of cabbage and lettuce turn yellow before the young inner leaves, as the plant withdraws nutrients and water from them in order to prolong the life of the central shoot. Rapid cooling and storing at high humidity will delay the deterioration of the leafy tissue.

It is useful to remember that ethylene is a gas produced in minute amounts by ripening fruits and acts as a ripening regulator. It also accelerates the senescence of green vegetables with which it comes into contact. These two groups of produce must therefore not be stored near to each other in a confined space. A bag of apples, or pears or tomatoes, for example, could produce enough ethylene gas to affect adversely leafy vegetables if stored for several weeks in a refrigerator. However ethylene would not build up to active concentrations in a well-ventilated shed.

Root crops

To keep in good condition, carrots, parsnips, beetroots, swedes and turnips require low temperature and the highest possible humidity. For these crops storage can often be avoided simply by leaving the roots in the ground until required for the kitchen (see Fig. 3.1). In this way really fresh carrots and parsnips can be lifted throughout the winter and well into the spring.

Our common root crops are natural storage organs, accumulating sufficient food in the first season of growth to enable them to flower early in the second season. Crops grown for winter supplies should be sown early enough for them to mature by late autumn, but late enough to ensure that they do not mature too early, as unwanted late growth can lead to over-size roots and the undesirable toughening of existing tis-

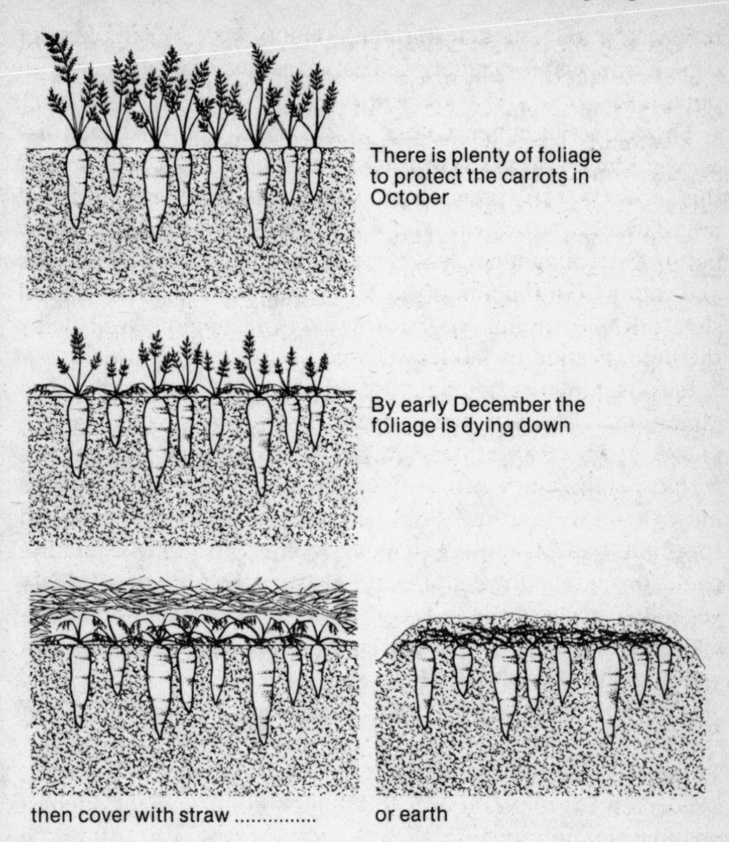

Fig. 3.1. Covering carrots for the winter.

sues. It is best to sow as late as recommended for any particular cultivar (for guidance see Chapter 2). Slow growth will continue during the winter until the plants are harvested, so they should be grown at a sufficiently high plant population to minimize the risk of over-size roots.

Freezing injury. The storage roots of our common root vegetables can withstand slight frosts (see Table 3.2) because the small amounts of ice formed are located harmlessly in between the cells of the outer tissues. The effects can be seen as small spaces or cracks in the tissues when the ice has thawed. The heat of respiration, which is difficult and expensive to get rid of in summer crops is, in this instance, valuable in keeping the tissues from freezing. However, when the storage roots are subjected to temperature lower than 26°F (–3°C), the amount of ice formed is too great to be contained within the root. Large masses are formed which break through the root surface and cells may be destroyed by ice forming in them. Rotting of the roots may follow as fungi and bacteria invade the cracks and the useful life of the vegetable will be shortened. Carrot cultivars of the Chantenay and Autumn King types are less susceptible to freezing than those of Amsterdam Forcing and Nantes types, a point to be considered when choosing cultivars for overwintering.

Frost protection. There are many cultivars of carrots in the Chantenay and Autumn King groups, and late parsnips, which grow very slowly during the winter and are ideal for complete overwintering. The carrot leaves form a canopy over the roots and help to protect them from damaging frosts. However, neither the leaves nor the roots can withstand a succession of

Table 3.2 Storage conditions for 'topped' root crops

	Ideal storage conditions			*Number of days of likely storage life if harvested and stored at low ambient temperature 40–50°F (4–10°C) in winter*
	Temperature and humidity	*Freezing-point*	*Likely maximum storage life (days)*	
Beetroot	40°F (4°C) highest possible humidity	30°F (–1°C)	120–200	120
Carrot	32°F (0°C)	30°F	60–150	90
Parsnip	highest	(–1°C)	150	60
Swede	possible		120	60
Turnip	humidity		150	60

heavy frosts (below 23°F (–5°C)) and by early December a large proportion of the leaves blacken and offer little protection to the roots. At this time additional protection must be given, but it is unwise to cover crops before early December as the extra insulation will tend to retain heat in the soil, and the crop is more likely to continue growth. It is a good idea to apply slug bait before covering.

The object of good insulation is to trap air but the material loses its effect if it becomes wet and matted. Suitable materials are loose straw, leaves, sand, sandy or peaty earth, peat, or polythene sheeting. Wheat straw is one of the best materials to use as it is not easily compacted over winter, is not easily blown away and is easily gathered and disposed of. Any remaining fragments can be dug in. An average bale will cover about 6 square yards to a depth of 6 inches or more, and this is usually sufficient to prevent carrot roots from freezing.

Parsnips are hardier and can survive most winters with very little extra protection; a light covering of the crowns with earth is usually adequate. Parsnips respond to low temperatures by converting some of their starch to sugar, which makes them taste sweeter. At least 8 in. of straw is needed for beetroot, swede and turnip, as the edible parts are partly above the ground.

The success of overwintering root crops in the ground can also be affected by the location and the type of soil. A sheltered spot is preferable to an exposed one, and a well-drained soil is essential as long periods of waterlogged conditions can lead to many disorders. Furthermore, a freely-draining soil has better insulating properties than a waterlogged one. Gardeners with heavy soil would find it worthwhile to prepare a special area for overwintering carrots by adding large quantities of sand or peat (or both). A raised bed will help drainage in the winter and poorly-drained areas should be improved by digging a channel to take the excess water away.

Loss of quality. The quality of beetroots, swedes and turnips tends to deteriorate more than that of carrots and parsnips when left in the ground, possibly because they are more exposed, being at or above soil level. It is normal to harvest these at some stage of the winter in all but the most sheltered districts. Common quality defects are excessive growth of the shoot or root, splitting, and the development of rough inedible tissue on the exposed surface of the root. Frequently the roots will become more fibrous in texture because of the deposition of cellulose thickening as this is part of the natural ageing process. This deterioration in quality can be minimized by lifting and storing indoors.

Harvesting and storage. Once lifted any remaining leaves act as wicks drawing moisture out of the root, so they are best removed by twisting them off. This simple method is adequate, as broken leaf-bases usually drop off after a few weeks, leaving a neat, protective scar. Twisting off the leaves of beetroot is much better than cutting, as the sap may be lost through the cut tissues for some days. Damage to the crown of these root crops at harvest must be avoided as the terminal shoot is vital to the well-being of the plant, for without it the root becomes moribund and will eventually die and decay.

Root crops should be stored so that there is some circulation of air. Undamaged roots can be roughly size-graded to facilitate packing and stored in wooden boxes (slatted wooden crates are ideal). The roots should be laid down in layers with enough sand between the layers to cover them. The sand separates the roots, reducing the chance of the spread of disease, and allows the movement of air and moisture into and out of the roots. Simpler storage methods, such as putting them in polythene bags, usually result in more waste from rotting because of the root-to-root contact. Clamps are not now generally recommended for root crops because they take a lot of time to construct and generally give poorer results than crops stored in the ground. The possible exception is beetroot, which will store reasonably well in polythene sacks and, for large quantities, in clamps.

The clamp should be made on level ground by piling up

beetroot so that the triangular cross-section of the clamp measures about 30–40 in. (75–100 cm) across the base and in height. The roots are covered with a 6 in. (15 cm) layer of straw, and this is followed by a 6 in. layer of soil put in position in December, when the roots are cool. Beetroot stored in this way will keep well until March or April.

Root crops grown in well-drained soil and lifted in dry conditions are usually clean, and need not be washed before being stored. Washing means extra handling and is likely to increase root damage and is therefore undesirable. If, however, roots are lifted from a heavy soil, perhaps in wet conditions so that little of the surface of the roots is visible, it is worth washing them to remove excess soil, which may carry a number of pests and diseases. If high levels of pest attack are suspected, washing will enable damaged roots to be detected and removed before they are stored.

Potatoes

Main crop potatoes are mature when the leaves die down naturally, and the tubers should be harvested before there is any risk of frost damage. They are best harvested in dry conditions so as to obtain the tubers as clean as possible. Once any damaged tubers have been removed for using up first, sound potatoes should be put into double-thickness paper sacks (obtainable from horticultural suppliers) which are then tied at the neck. Potatoes harvested under wet conditions should be surface-dried before storing. These paper sacks allow surplus moisture to escape and the moisture content of the air around the tubers will equilibrate at a suitable level. Paper sacks exclude low levels of light from the tubers. This is important because light stimulates the formation of poisonous alkaloids in the surface tissues as they turn green. This green tissue must be removed before cooking. Jute and hessian sacks are also suitable provided light is not allowed to reach the tubers. Polythene sacks are not suitable for storage as the humidity will rise and stimulate the development of sprouting as well as some disorders and diseases.

There is no hurry to reduce the temperature of the potatoes to below 50°F (10°C) after harvest. Indeed, temperatures between 50 and 60°F (10 and 15°C) are beneficial for a few weeks, as the development of a layer of protective corky tissue on the skin and the healing of any minor wounds are stimulated at these temperatures. It is, of course, normal to store potatoes at prevailing (ambient) temperatures in a cool, frost-free place. Large quantities of potatoes may warrant the construction of a field clamp, using the same method as described for beetroot, but omitting the layer of straw. Storage for lengthy periods below 40°F (4°C) encourages the undesirable sweetening of the tubers, which is detectable after cooking. This is particularly noticeable with potatoes for chipping or roasting as the sugars formed at low temperatures caramelize on cooking and this results in very dark chips or roast potatoes which can be somewhat bitter. Low temperature effects can be avoided by giving extra insulation, such as layers of newspapers, when the weather is very frosty. Wooden boards under the sacks will insulate the potatoes from below.

The rate of water loss from potatoes is surprisingly low in paper sacks but this will increase as sprouts appear on the tubers in the spring. The sprouts act as wicks drawing moisture out of the tubers and will make the potatoes spongy in time. Sprouting is also accompanied by sweetening of the tuber as food reserves are mobilized in readiness for regrowth. A few extra weeks storage life can be obtained if the sprouts are rubbed off.

Onions

Ideal conditions for storing onion bulbs are a combination of low temperature 32°F (0°C) and low humidity but these conditions are difficult and expensive to provide and many months of storage life can be obtained at the low, ambient temperatures normally found in an unheated building. The storage procedure for bulbs grown from sets or from seed is similar except that bulbs from sets are harvested 2–3 weeks earlier and have a shorter storage life.

Harvesting and drying. Complete drying-off of the leaves to give a well-sealed neck is essential for successful storage. In early September the leaves fall over and die down indicating that the bulb has matured. At this time sprouting-inhibitors produced in the leaves are translocated to the bulb, and for this reason leaves must be left intact and dried on the bulb. When the leaves have fallen over in the majority of the plants, or in any case by mid-September, they should be lifted to encourage drying-off. If the weather is good the bulbs can be dried-off in the garden, laid out in rows, for a week or so. Drying is complete when all the green tissue and papery skins around the bulbs are 'rustling' dry. The roots which are fleshy at lifting time should also have withered. In a cool season, a proportion of the plants will develop extra leaves and have a thick, fleshy neck. These plants will not die down and store well, so should be used up as soon as possible. Drying is encouraged by warm, airy conditions and the bulbs should be laid out to dry under cover if the conditions for drying outside are poor. The staging of an empty greenhouse is ideal for this purpose. Any warm place can be used to assist drying, but temperatures over 80°F (27°C) should be avoided because of the increased risk of splitting the skins which are necessary for successful long-term storage. The higher the temperature the darker the colour of the skin becomes. This looks attractive but does not improve storage performance!

Contact of the bulbs with wet soil or moisture during drying causes staining of the skins. Thus, extended outdoor drying (with a strong possibility of rain or dew) often results in heavily stained skins, whereas bulbs removed and dried under cover should be virtually unblemished. However, staining does not affect the eating quality of the bulbs.

Storage. Onion bulbs must be stored under cover with a good circulation of air. Slatted-trays containing not more than two layers of bulbs are ideal for this purpose. Alternatively, the bulbs can be put into nets or made into 'ropes' for hanging in a draughty place, for there is no danger of desiccating them by over-ventilation. A good circulation of air will remove moisture produced continuously by respiration and will keep the skins 'rustling' dry. In well-dried bulbs the pale, shiny tips of new roots can be found, with a little searching, in a circle around the old withered roots. The new roots should remain about the size of a pinhead (1 mm) until January when new growth may begin. Any premature growth of the new roots before this is a warning that dormancy has been disturbed and that sprouting could be on the way. If water droplets come into contact with the base of the bulb for more than a few days, new root growth will occur any time after the bulbs are lifted. This may happen if drying outside was poor because of prolonged rain, or if bulbs are put into polythene bags so that condensation builds up. It can be seen that it is essential to keep water away from onions in store.

Until late December the shoot inside the bulb should remain white, but in January it begins to turn yellow as the process of sprouting develops. New root growth also coincides with the development of the shoot. The time taken for the emergence of the sprout from the bulb depends on the storage temperature – the lower the temperature, the better. There is little danger of damaging the bulbs by frost as they have a low freezing-point, for they even recover from being frozen at about 26°F (–3°C) with little apparent damage. If they do freeze it is best to leave them undisturbed to thaw slowly when the ambient temperature rises. Over 50 per cent of the bulbs will start to sprout from mid-March to the end of April. Storage can be prolonged after this time in a refrigerator which can provide the ideal conditions of low temperature and low humidity. Bulbs should be placed in a net in order to benefit fully from the conditions. If onions are to be stored in a refrigerator, they must be placed there before mid-January or before yellowing of the internal sprout starts. If, despite all your efforts, sprouting occurs prematurely, there is the minor consolation that the sprouts are quite edible but the bulbs should be used up quickly.

Bulbs from the overwintered crop are not usually stored for

any length of time. However, they can be dried and stored at ambient temperatures in the same way as the main crop, but with bulbs harvested in June and July, 50 per cent will have sprouted by December.

Vegetables susceptible to 'chilling injury'

The ancestors of certain beans, cucumbers, marrow, pumpkin and other cucurbits, sweet peppers and tomatoes all came from sub-tropical areas of Central and South America. Consequently the plants, including the fruit, are poorly-adapted to temperatures below about 50°F (10°C) (see Table 3.3) although the seed can survive such temperatures. At temperatures below 50°F (10°C) these fruits develop pitting of the surface, loss of flavour and texture, and sometimes internal discolouration. If chilling persists for long enough, internal tissues become very soft and have a water-soaked appearance, and their breakdown is completed by spoilage organisms.

Table 3.3 Storage conditions for vegetables susceptible to chilling injury

	Storage conditions		
	Temperature and humidity	*Freezing-point*	*Likely maximum storage life (days)*
Beans, French and Runner	All require 41–45°F (5–7°C) and high humidity	All freeze at 30°F (–1°C)	8
Cucumber and Courgette			14
Marrow and Pumpkin			60
Sweet Peppers			21

The effects of chilling injury are cumulative; a short period at 34°F (1°C) is as bad as a longer period at 40°F (4°C). Tomatoes and cucumbers kept for a week in a refrigerator at about 40°F (4°C) will be irreparably damaged and will be of poor flavour and texture. Furthermore, cool nights in August and September can initiate chilling injury in the garden even before harvest. These fruits and vegetables are best stored in a cool cupboard or room, partly but not completely wrapped, to allow some circulation of air. If they are to be kept for only a few days refrigeration is acceptable as chilling injury symptoms will not have time to develop, though the flavour will probably suffer.

With potatoes and beetroot chilling injury results in an increase in decay and loss of quality at temperatures near to freezing-point.

In conclusion, much can be done to extend the season of supply of many vegetables by simple methods of storage. Neither expensive equipment nor complicated techniques are required to prolong the availability of your own vegetables for weeks or months.

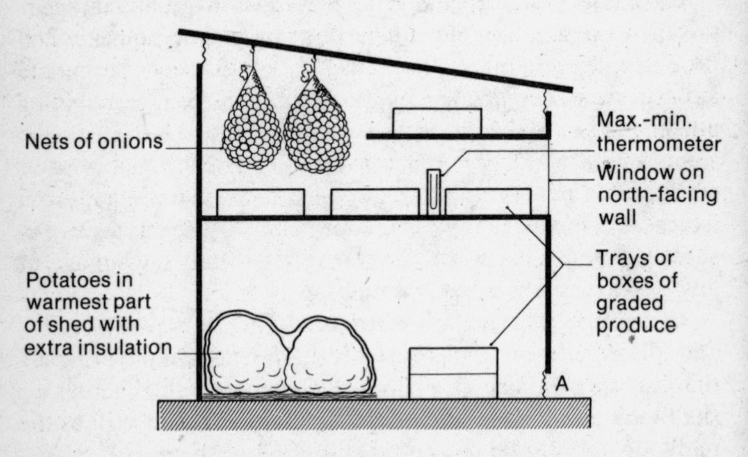

Fig. 3.2. Cross section of a garden shed adapted for the storage of produce. Air circulates in the shed, entering through fresh-air vent A and leaving via small gaps between eaves.

4 Weeds and weed control

The most usual definition of a weed is 'a plant out of place' or 'a plant growing where it is not wanted'. On this basis any kind of plant can be a weed, even a cabbage or a lettuce, for example, if it happens to be growing in an onion bed. However, we normally reserve the term for wild plants like groundsel or ground elder which establish themselves without our help and which keep on appearing from year to year despite efforts to get rid of them.

The common weeds of vegetable gardens are adapted to survive in cultivated ground. Some do it by growing quickly and producing large numbers of seeds. Not all the seeds germinate at once; many remain alive but dormant in the soil, germinating and giving rise to seedlings only when the ground is disturbed at some future time. Others survive by forming persistent roots or stems, often deep in the soil, in which food is stored and from which new shoots can arise in later years. These strategies ensure that weeds are always there, ready to take advantage of any opportunity to grow and multiply.

Agriculture has been described as 'a controversy with weeds', and this is no less true of vegetable gardening. On the principle that it is a wise approach to know the enemy, in this chapter we shall look at some of the characteristics of weeds as well as the methods that can be used to keep on top of them.

WHY ARE WEEDS HARMFUL?

The purpose in having a vegetable plot is to grow good vegetables, preferably with a minimum of effort, and if weeds make it more difficult to achieve this, then their presence must be considered undesirable. Weeds interfere with crop production both directly and indirectly. Their main direct effect is to reduce crop growth and yield by preventing the crop from using resources that would otherwise be available to it. Indirectly, they can affect the crop by harbouring pests and diseases or they can simply get in the way and make it more difficult to harvest the produce. In order to avoid these effects something has to be done, and this involves time and energy. If there are only a very few weeds they may present no immediate problem, but left untouched they will certainly multiply and become an increasing nuisance in the future.

Direct effects

It goes without saying that if we grow vegetables on weed-infested ground and take no steps at all to deal with the weeds, the crops will be unable to realize their full potential. This is primarily because of competition. The weeds will take up water and mineral nutrients from the soil so that less are available for the crop, and they may grow above it and prevent the sunlight from reaching it. How serious these effects are depends on various factors.

Some crops are better able to resist competition from weeds than others. A lot depends on the size of the seedling, how quickly it grows and how soon the crop covers the ground. Onions are one of the crops most susceptible to competition. Germination and early growth are slow and the leaves of the young plants are slender and upright, so that quick-growing broad-leaved weeds have a distinct advantage. Broad bean, on the other hand, produces a large seedling so that the advantage is with the crop. If weeds are allowed to remain right through the season in a crop of onions grown from seed, almost the entire crop will be lost; with broad beans, the effects will be much less pronounced and may only be severe if there is a shortage of water in the soil at the time when the pods are filling out. Transplanted crops, especially those in soil blocks or planted out from pots, also have an advantage over weeds germinating from seed.

Weeds themselves differ in competitive ability, although this can vary with the circumstances and the time of year. On the whole it is the tall-growing, vigorous weeds like fat-hen which have the most pronounced effects; seedlings of fat-hen which appear in late summer, however, only form small plants, although these still produce plenty of seeds. Chickweed, generally a low-growing plant, may not present much of a problem in a tall crop during summer, but plants which have overwintered grow rapidly on a fertile soil in early spring and can completely smother crops like salad onions and spring cabbage.

The fact that weeds are present in a crop does not necessarily mean that competition is actually taking place. During the early stages of growth, when both crop and weeds are still small seedlings, each can obtain all the resources needed to grow without interfering with any of the others. As the seedlings increase in size, however, their root systems explore greater volumes of soil and when the supply of nutrients or water is no longer great enough to meet the needs of all of them, competition begins. The weeds must be removed before this happens, otherwise there will be detrimental effects on crop growth.

It is not easy to pinpoint the time at which competition begins, since it depends on several factors. It will occur sooner, for example, when the temperatures are high and rapid growth is taking place than it will in very early spring when things are happening slowly. It is also influenced by the kinds of weeds present and by how many of them there are, as well as by which crop it is. If there are only very few weeds, no more than one or two per square yard (metre), they may have no detectable effect on the crop even if they are left until harvest, but the more weeds there are the shorter the period before competition starts. Experiments with several different spring-sown vegetable crops and naturally-occurring weed populations showed that weeds and crop can co-exist without interference for about three weeks after the crop has come up. This is a reasonable figure to take as a guide.

Once competition does begin, the effects can be quite startling. In some research with onions, the weeds were carefully removed from separate plots at different times during the growth of the crop and the plots were then kept clean. It was found that after competition had started, the final yield of bulbs was being reduced at a rate equivalent to almost 4 per cent per day. So that by delaying weeding for another fortnight, the yield was cut to less than half that produced on ground kept clean all the time. In these experiments there were many more weeds per unit area than there were crop plants, which is what usually happens, and they also grew more rapidly than the crop. By early June the weight of weeds per unit area was more than twenty times that of the crop, and the weeds had already taken from the soil about half the nitrogen and a third of the potash which had been applied in the base fertilizer dressing.

Another reason why onions are particularly susceptible to weed competition is because they only produce a limited number of leaves before they start to bulb (see p. 175). So if the weeds are removed, there is little recovery from the ill effects that may already have occurred. Other crops, like red beet for example, keep on producing new leaves and have much better powers of recovery after weeding, even when this has been delayed. Once a crop has become successfully established and has formed a good leaf canopy, any further weed seedlings that appear are usually suppressed; the crop has the advantage and the weeds are unable to compete effectively. Seeds of many weeds need light for germination, and there is evidence that the presence of a leaf canopy can actually prevent germination. This is because the quality of the light filtering through green leaves is altered in such a way that it tends to promote rather than overcome dormancy in seeds present at or very close to the soil surface.

So far we have been considering the effects of weeds developing from seeds, stimulated to germinate by the soil disturbance involved in preparing the seedbed for the crop. Weed shoots which come from perennials with persistent underground root systems have the advantage of being able to use the stored food and can grow very quickly. Although they may not compete with the crop directly while it is still in an early stage of growth, since the root systems will be at different levels in the soil, they can soon overtop the crop and shade it. There is evidence for some kinds of perennial weeds that substances produced by either the living or decaying parts of the plants can be released into the soil and may inhibit the growth of other plants – a phenomenon called allelopathy. In practice, however, it is difficult to assess these effects and to distinguish them from straightforward competition for limited resources.

Indirect effects

The fact that weeds serve as hosts for many of the pests and diseases that attack crops is often quoted as a reason why they should be eliminated from the vegetable garden. It is perfectly true that they do harbour a wide range of organisms including insects, fungi, viruses and plant parasitic eelworms, and that these may be able to multiply either on the weeds or within their tissues. The question is whether this really matters from the practical point of view.

So far as insect pests are concerned, the evidence suggests that weeds probably do not play a very important role in the vegetable garden. Although aphids, for example, may be found infesting particular weeds, they are not necessarily the same species or strains that attack the crops. Tests with the common cruciferous weeds, such as shepherd's purse, have shown that they do not act as hosts for the larvae of the cabbage root fly. One exception was found to be wild radish, but this is not usually a weed of established vegetable gardens. It is true that some pests, for example cutworms, may first attack weeds and then move onto crops, but in general it seems that the importance of weeds as hosts for insect pests of vegetables on a garden scale has been over-emphasized. Crop residues, on the other hand, may be of much greater significance. Carrots left in the ground over winter can certainly increase the risk of serious carrot-fly attack in the following year.

A similar conclusion applies in relation to the nematodes (eelworms) that attack vegetables. The presence of black nightshade would be likely to keep up the numbers of potato-cyst eelworm while onion-stem eelworm can carry over on a number of common annual weeds, but there is little evidence to show that weeds on the vegetable plot are of any importance in this respect. Again, with the fungi that cause diseases in vegetables, weeds probably do not play any important role. Fungi exist in different strains, and those found on weeds may not be the same ones that attack the crops. The 'bridging' effect of having crops of a particular kind present throughout the year is probably a more significant factor. The organism which causes club-root of brassicas can survive in the living and dead roots of weeds in the same plant family, and in this instance, their presence can be detrimental in reducing the benefits of crop rotation.

In contrast, the role of weeds in providing a reservoir and source of infection for viruses which cause diseases in vegetables has been underestimated, partly because the infected weed plants often show no symptoms themselves. Overwintering weeds like chickweed and groundsel frequently carry cucumber mosaic virus which may then be transmitted by aphids to a range of crops, including marrows and tomatoes. Chickweed, which shows no symptoms, is especially significant as a source of infection because it has been shown that this particular virus can persist in the seeds produced by an infected chickweed plant. These can then give rise to more infected chickweed plants at some later time. Beet western yellows virus has become increasingly troublesome in lettuce in recent years, causing characteristic yellowing of the outer leaves as the plants develop. This virus also occurs in common weeds like shepherd's purse, groundsel and hairy bittercress, and if these are not destroyed in winter they serve as a bridge and provide a source of infection for lettuce grown in spring. Shepherd's purse and most other cruciferous weeds carry cauliflower mosaic virus, and there are others which could be cited. There are, therefore, good reasons for ensuring that the vegetable garden is cleaned up at some stage during late autumn or winter in order to prevent carry-over of viruses as far as possible. Weeds, however, are not the only sources of infection; many ornamental plants grown in flower borders serve as hosts from which vegetables can acquire viruses transmitted by aphids. This is true of delphiniums, for example, while wallflowers provide an effective bridge whereby turnip mosaic virus can be carried over the winter period.

There are some other indirect effects of weeds which can be very important in commercial vegetable production. When crops are harvested mechanically the mere presence of weed vegetation, even at a level such that there is no competition with the crop, can slow down the operation and make it much more costly. Berries and seed heads can contaminate the harvested crop and make it unacceptable for processing. So it is hardly surprising that the commercial vegetable grower has to maintain a high standard of weed control. Even in the garden twining weeds like field bindweed or black bindweed can make it difficult to pick a crop, while the presence of annual nettle makes the harvesting of one's own produce unpleasant. The presence of weed vegetation may also tend to promote the spread of fungal diseases by impeding airflow and maintaining a moist atmosphere close to the crop.

ANNUAL WEEDS

These are weeds which develop from seeds and which grow, flower, set seed, and die all within a year or less. Even among annuals there are differences in the time taken to complete the life-cycle. The frequent cultivations associated with growing a rotation of vegetables in a garden tend to discourage those which have a relatively long life-cycle, and such common cornfield weeds as charlock and wild radish are usually not a problem in gardens. The ones that flourish are those which take only a short time to reach maturity and produce more seeds again. Chickweed, groundsel, shepherd's purse, annual meadow grass and field speedwell are among those which are universally present in vegetable gardens, and some of them can often complete two and sometimes more generations within the year.

Growth habits

Annual weeds exhibit great variety in growth form. Some of them, like fat-hen and annual nettle are erect plants which grow straight upwards. If they have plenty of space they branch out sideways as well and become very large; if they are crowded they just produce single stems. Annual nettle starts to shed seeds while the plant is still small – little more than four weeks after the seedling has emerged – and it goes on seeding right through the season until the winter frosts. Fat-hen, however, sheds all its seeds at more or less the same time at the end of the summer.

Quite a few of the common weeds, like chickweed, knotgrass and the speedwells, have a prostrate, creeping habit. They branch extensively and spread over the ground, forming a mat. These weeds tend to produce seeds throughout most of their life; ripe seeds are shed from the capsules formed at the base of the branches while more flowers continue to be formed at the tips. Yet other annual weeds, among them shepherd's purse, hairy bittercress and the mayweeds, begin by forming a flat whorl or rosette of leaves and then later on the flowering stem develops from the centre of the rosette. These weeds usually produce all their seeds at much the same time, but how long they remain in the rosette stage depends on the time of year at which they germinate. Plants appearing in spring and early summer go up to flower quickly, whereas if they emerge in late summer or autumn they may stay as rosettes through the winter and not flower until the following spring. Such a pattern of behaviour approaches that of biennial weeds. These usually germinate in spring and the plants remain in the vegetative stage during the first season while they build up food reserves; flowering and seeding then take place in the following year. Biennials are common in hedgerows and waste places but again because of the frequent cultivations, they are not usually a problem in a vegetable garden. Two which do sometimes occur are spear thistle and white campion.

Many common annual weeds of gardens are perfectly frost-hardy and can continue to make some growth even at low temperatures, while some of them like the speedwells and red deadnettle can be found in flower in winter. Others normally die off during winter and then fresh seedlings come up in spring; black nightshade is one which is killed by the first severe frost. There are also differences in the tolerance of weeds to drought. Annual meadow grass and chickweed, with their fibrous root systems, are among the first to be killed as the soil dries out whereas the deeper-rooting fat-hen and black nightshade are able to keep going for much longer in dry summers like that of 1976.

Seed survival

As already mentioned, one of the important characteristics of annual weeds is that they produce large numbers of seeds. The numbers may be very large indeed; a single big plant of fat-hen, for example, may produce 70,000 seeds. However, the average numbers per plant are much less, because there are usually a lot of small plants present. In fact another important feature of annual weeds is their ability to produce seeds even when the plant size is very much reduced through lack of resources. Tiny plants of annual meadow grass growing on an ash path, for example, may still have one or two seeds each even though they are scarcely visible. It is the small plants which are so easily overlooked.

Measurements of the numbers of seeds shed onto the soil where crops have been allowed to become weedy show that there can quite easily be 5,000 per square foot (54,000 per square metre). Many of these, of course, would probably be eaten by birds, and others may germinate immediately if the soil is moist. It is common to see a dense carpet of seedlings beneath a big plant of annual nettle, for example, and large numbers of seedlings of groundsel and smooth sowthistle often appear in late summer wherever plants of these weeds have been allowed to mature. Even with weeds like these, not all the seeds will find suitable conditions for rapid germination and some will become incorporated into the soil. Moreover, seeds of many weeds have a built-in dormancy mechanism which prevents immediate germination or restricts it to only a small proportion of the total number. Seeds of fat-hen and knotgrass are unable to germinate when they are shed from the parent plant in summer or autumn, and will not do so until they have experienced a period of exposure to low temperatures during winter.

As a result, there is almost invariably a sizeable 'seed bank' in the soil made up of seeds which are alive but dormant. Studies of soil samples taken from fields cropped commercially with vegetables have shown that it is quite common to find 900 living seeds per square foot (10,000 per square metre) in the top 6 in. (15 cm) of soil. In very weedy fields as many as 7,000 per square foot (75,000 per square metre) have been recorded. There are no similar figures for the numbers of weed seeds present in garden soils, but the situation is probably very much the same.

The action of disturbing the soil to produce a seedbed causes some of the seeds present in the soil to germinate, and a flush of seedlings appears unless it happens to be too cold or too dry. It is still not known precisely why soil disturbance has this effect. It used to be thought that it was largely a question of aeration of the soil, but it now seems that exposure of the seeds to light is probably a major factor. However, not all the seeds will be affected; some will retain their physiological dormancy while others may not encounter the right conditions at the time so that they too remain dormant for the present. Recent research has shown that the proportion of seeds in the soil which gives rise to seedlings after any one cultivation is only small. Even under favourable conditions, in spring, say, when the soil is moist, probably no more than one-twentieth of the total store of seeds will germinate so that the weeds which actually appear in a crop represent only the tip of the iceberg.

Besides the seeds which make it to the surface and establish as seedlings, others will be stimulated to germinate but will die because they use up their food reserves before the seedling shoot gets there. Moreover, other seeds present in the soil will die without ever germinating. So that provided no fresh seeds are allowed to come in, the seed bank will tend to decrease naturally. Research has been carried out to determine how quickly this takes place, and the results have proved both interesting and useful. It was found that with a soil which was carrying vegetable crops and which was frequently cultivated, about half the seeds were lost from one cause or another during the year. This means that if no fresh seeding were allowed, the seed bank would be halved each year and would reach a level of 1 per cent of its initial size after seven years. So the old saying 'one year's seeding, seven years' weeding' is perhaps not far from the truth.

There is really no way of speeding up the process and getting the seed bank to decline more rapidly. Cultivating the soil more frequently than is necessary to grow the crops properly will not have very much additional effect. One hope for the future is that a chemical might be discovered which could be applied to the soil and which would stimulate all the seeds to germinate at once so that they could be killed. Certainly there are growth-regulator chemicals which on a laboratory scale are very effective in bringing about germination of dormant seeds of particular kinds of weeds. One problem is that not all seeds respond to any one chemical, and at present this approach has not met with much success in practical terms. The opposite approach would be not to try to make the seeds germinate, but to kill them while they are still dormant. Several of the soil partial sterilant (fumigant) chemicals used in commercial vegetable production, primarily for disease or eelworm control, do have this effect, and the benefit they give in terms of weed control is a useful bonus for the grower. For the gardener, however, the important thing is to prevent the weeds that do come up from reaching maturity and adding their quota of fresh seeds to those already in the soil as a result of seeding in previous years. If this can be done, the seed bank will progressively decline through natural causes, and life will become that much easier.

The question 'how long do weed seeds live?' is one which is often asked but has no simple answer. For one thing, there are differences between species; seeds of groundsel, for example, are relatively short-lived compared with those of fat-hen and many other species. However, there are some general principles. Survival will be shortest when the seeds are close to the soil surface; here they are exposed to light and to the greatest fluctuations in temperature, both factors which promote germination. If the seeds are initially distributed throughout the topsoil, say, to a spade's depth, then regular cultivation will keep bringing seeds near the surface and give them maximum opportunity to germinate. If the soil remains undisturbed, the life of the seeds is prolonged, and this is especially true of seeds which lie some distance below the surface. When buried deeply, seeds of various weeds, among them chickweed, have been shown to survive for forty years or more, and there is evidence that in a few species a proportion of the seeds can remain alive for much longer than that.

Time of emergence

In general, the greatest emergence of weed seedlings takes place either in spring as the soil warms up, or in late summer/ early autumn. In winter the temperature is too low for germination of some seeds to begin, and in any case all growth processes are slowed down during the cold weather. During the height of summer it may be too hot for germination of certain species, but more commonly it is lack of soil moisture which prevents germination at this time. As already emphasized, seeds freshly shed from the parent plants may germinate immediately. It is possible, however, to give some indication of when seeds present in the soil are most likely to produce seedlings, and this is done in Fig. 4.1 for twelve annual weeds.

Quite a number of the commonest weeds are able to germinate over a wide range of temperatures, like the first five species in Fig. 4.1. They are able to take advantage of any opportunity to produce seedlings, and the seedlings can survive no matter when they emerge. This obviously contributes to their success as weeds in vegetable gardens. Others, like fat-hen and annual nettle, have their peak period of emergence in spring but few seedlings appear after early autumn.

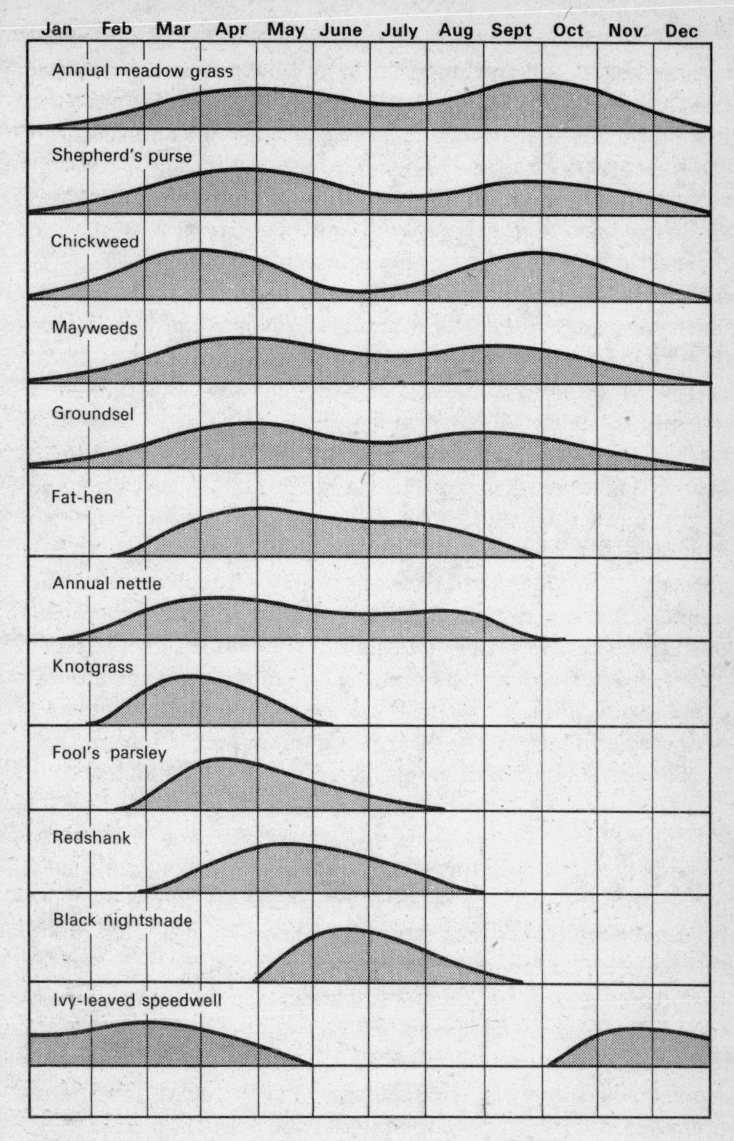

Fig. 4.1. Times of year at which seedlings of twelve annual weeds are most likely to appear following cultivation.

Black nightshade is one of the species with a restricted period of emergence, not starting until early May and stopping again in September. This means that seedlings often appear after the first flush of spring weeds has been dealt with, and they can easily be overlooked.

Some weeds have specialized means of restricting the period of emergence so that seedlings only appear at times when there is a chance of being able to complete the life-cycle. Research has shown that seeds of knotgrass need exposure to low temperatures in order to overcome dormancy, and emergence then begins in early spring (Fig. 4.1). High temperatures at the end of May then re-impose dormancy and cause germination to stop, so that the seeds still present in the soil require a further exposure to winter cold before they again become capable of germinating. Ivy-leaved speedwell behaves in the opposite way. It is the high temperatures of summer which overcome the dormancy, and seedling emergence does not begin until October. Seeds of this weed germinate when the temperature is fairly low and seedling emergence continues during winter and spring. During this time dormancy increases in those seeds which remain in the soil, and this is again overcome during summer. With these weeds, therefore, there is a definite seasonal pattern of seedling emergence, and the time of their appearance can be predicted with some confidence.

Mechanical control

Vegetable crops are, for a variety of good reasons, normally grown in rows and it is necessary to deal with weeds growing both between and within the rows. The possibilities for doing this by physical means come down to four: the weeds can be cut off, pulled up, smothered or burned off. Where there is no crop, digging is an additional option.

Hoeing. This is the traditional, and very effective, method of removing annual weeds growing between the crop rows. The usual tool employed is the Dutch hoe, an angled blade on the end of a long handle. This blade is moved back and forth parallel to the soil surface but just beneath it, so that the weed seedlings are cut off below ground level. It is important to keep the blade sharp so that it cuts cleanly and also to keep the depth of working shallow. Hoeing too deeply serves no useful purpose, and indeed is counter-productive since it brings more weed seeds to the surface where they can germinate and also leads to losses of soil moisture. It is also likely to cause damage to the root systems of the crop seedlings. The best time to hoe is when most of the seedling weeds have emerged and the crop rows are clearly visible. Hoeing earlier than this means that it will soon have to be repeated since a lot of the weeds will still be germinating beneath the surface; moreover, there is a risk of cutting into the crop row. Delaying too long, on the other hand, may allow the weeds to begin to compete with the crop. They will also be more difficult to cut through, and may quite easily root again if it rains. Ideally, hoeing is best done in a dry, sunny period but while there is still moisture just below the surface. The blade then flows easily through the soil and accurate working is possible because it does not bounce off dry clods and dislodge crop seedlings, while the weeds that are hoed off quickly die.

Besides the traditional pattern of Dutch hoe, there are nowadays various hoes of improved design, often with a double-acting blade. In larger gardens a wheeled hoe, equipped with L-shaped blades which can be set for different row distances, can be very useful in speeding up the work. With a friable soil and straight rows, very accurate working is possible. Motorized tools can also be used, either with blades or with tines that spin out the weed seedlings. The other common type of hand hoe, the draw hoe, is really not very suitable for removing seedling weeds from between rows of crops. Because it is used with a chopping action, it is easy to dig into the soil too deeply and bring up fresh seeds. It is also easy to leave some weeds undamaged, and because, unlike the Dutch hoe, its use involves walking forwards over the hoed ground, weeds not actually cut through can be trodden into the soil and may root again. Its main use in weed control is to deal with any larger weeds that survive until a later stage.

Hoeing is basically a quick and effective method of destroying annual weeds. There is no advantage in doing it more often than is essential to achieve this purpose; too-frequent hoeing may not only be unnecessary, but it can have detrimental effects as well (see p. 165).

Hand pulling. Hoeing is obviously a lot quicker than grasping each weed seedling individually and pulling it out. So it is an advantage to hoe as close as possible to the crop rows without causing any damage, and this is greatly helped by careful sowing in straight rows. Even so, there will be some weeds that appear actually within the rows among the crop seedlings, which must be removed by hand. Again, there is an optimum time for doing this. If done too early the weeds are difficult to get hold of and they are sometimes difficult to distinguish from the crop seedlings. Moreover, the process will probably have to be repeated very soon because more weed seedlings will have emerged. If weeding is left too late, competition with the crop may already have begun and it is also very likely that in pulling up the weeds the crop plants will also be disturbed. Weeds in the seedling stage, when they have still got only two true leaves, are usually easy to remove but it becomes more difficult as their root systems develop. There are some weeds, like fumitory, which are easy to pull up without disturbing the soil even when they are fully grown. Others, such as black nightshade, develop a forked root system which anchors the plant firmly and makes it very resistant to pulling; the stem can easily break off, leaving the plant to regenerate from branches produced at the base. Others again, like chickweed and annual meadow grass, produce a dense system of fibrous roots and tend to pull up with a mass of soil which can easily bring crop plants with it. Weeds like these root again very readily if they are left lying on the soil in moist conditions.

Mulching. Most annual weeds of gardens have small seeds which contain limited reserves of food and it is only those germinating within 1 or 2 in. (2.5 or 5 cm) of the surface that have sufficient reserves to enable the shoot to reach it. If the ground is covered with a mulch, an additional layer of something which keeps out the light, then the seedlings will not survive. Mulches can be applied between the crop rows and around transplanted crops and have the additional advantage of conserving soil moisture, especially if applied after rainfall or watering. The best organic materials include peat and various kinds of compost, such as weed-free garden compost or spent mushroom compost. Other materials which can be used include straw and grass mowings, though these have some disadvantages. Straw often contains weed seeds and there are usually some cereal grains which quickly germinate. Grass mowings may have a lot of seeds of annual meadow grass, which is a weed very common in lawns. It is, of course, important not to use mowings from a lawn recently treated with a selective herbicide since this could lead to crop damage. There is also a possibility that mulches of this kind can provide shelter for pests such as slugs. A great virtue of organic mulches is that when they have served their purpose they become incorporated in the soil and enrich it, though the nitrogen required in the humus-forming process may need to be replaced. Maintenance of a surface layer of compost is an essential feature of the 'no-digging' approach to vegetable growing (see p. 161). From the point of view of minimizing weed problems this system has the advantage that fresh weed seeds are not constantly brought up from lower down in the soil. Any seedlings that do appear in the surface layer can readily be dealt with because of its friable nature.

An alternative to organic mulches is to use strips of black polythene, laid on the ground and anchored at the edges with soil. Transplants, like brassicas, can be planted through holes made in the plastic. Large seeds of crops like sweet corn can also be dibbled in through holes, and potatoes can be grown in this way. Black plastic does not have the same advantage as clear plastic in raising the soil temperature but it does conserve soil moisture and unlike clear plastic, it keeps the weeds down.

Some suitable arrangement, such as slits in the furrows, is needed to let the water through to the soil. The disadvantages are the initial cost, the fact that the plastic has to be removed eventually, and the fact that to many people the appearance of plastic-covered soil is aesthetically unpleasing.

Flame-weeding. There are various proprietary flame guns available which can be used to burn off weeds from areas of uncropped ground. Some types can also be used for inter-row weeding among crops, although great care is needed to ensure that the crop is not damaged as well. Since the soil is not disturbed there is no stimulation of fresh germination, and dormant seeds which are on the soil surface or very close to it will be killed.

Digging. This is obviously not applicable to ground which is carrying growing crops. However, in parts of the garden from which crops have been cleared in autumn it is the traditional method of starting the preparation for next year's cropping. Annual weeds which are dug in are effectively killed, although burial must be complete; if any parts of creeping or tufted weeds remain above ground they can go on growing through the winter. Undoubtedly the spade is the best tool for the purpose except on heavy soils where it may be easier to use a flat-tined digging fork. To most people, there is something very satisfying in the contemplation of a piece of skilfully dug soil. If, however, there are seeds present on the weeds that are dug in it may be a question of postponing the problem: burial is the best way to ensure survival of weed seeds.

Chemical control

In the smaller vegetable gardens at any rate, the traditional methods of hoeing and hand pulling are still the mainstay of weed control. Where vegetables are grown commercially on large areas, of course, this is nowadays not feasible and growers rely to a large extent on selective herbicides for controlling weeds. These are chemicals which are applied either to the soil or over the crop after it has come up and which, for various reasons, act selectively to kill the weeds without harming the crop. The development of these techniques over the last thirty years has been a considerable success story, and without them it is difficult to see how vegetables could still be produced economically. To use them successfully requires a lot of care and know-how. Individual herbicides can only be used on particular crops, the application rate has to be carefully controlled, and such factors as the type of soil, the exact stage of crop growth and the weather conditions have to be taken into account. It is not surprising, therefore, that these techniques do not translate readily to a garden scale.

Nevertheless, there are a few herbicides which are available on the retail market and which can be helpful in the vegetable plot. It is most important to read the labels carefully before use, and to use them *only* on those crops for which they are recommended. The following paragraphs are intended to indicate the possibilities; they are *not* a substitute for the manufacturers' directions, which must always be followed.

Paraquat/diquat. This is available as a soluble solid formulation. It is not a selective herbicide and will kill the green tissues of both crops and weeds if it comes into contact with them. It acts rapidly, especially in warm weather, and even if rain falls shortly after application its effectiveness is not reduced. Once it reaches the soil it is inactivated, so that it cannot harm the roots of crop plants.

Paraquat/diquat has two main uses in the vegetable plot. First, it can be applied between crop rows using a dribble bar attached to a plastic watering can (Fig. 4.2.). Great care is needed in doing this, because if any of the solution splashes onto the crop plants they will certainly be injured and perhaps killed. Used in this way it serves as a 'chemical hoe'. The advantage over hoeing is that because the soil is not disturbed, few fresh weed seeds are caused to germinate and the effect therefore lasts longer. The second use is for killing weeds that have appeared on ground which is not actually being cropped

Fig. 4.2. Applying paraquat/diquat with a dribble bar to kill weeds between the rows of a transplanted crop.

at the time. On parts of the garden from which crops have been cleared in early autumn large numbers of seedling weeds often appear, and these will produce quantities of seeds before the time arrives for digging. This can be prevented by applying paraquat/diquat, either with a dribble bar or through a rose on a can. Shoots of any perennial weeds present will be killed but paraquat/diquat will have no effect on the underground parts, so the benefit is only temporary so far as these weeds are concerned. It can also be very useful for treating weeds that appear on ground which has been prepared some time before it is actually required for sowing or planting. Seeds can be sown immediately after application and plants put in after twenty-four hours. Besides being a lot quicker than cultivation in these circumstances, there are other advantages. The main flush of weed seedlings that follows soil disturbance is avoided, and if the soil is moved as little as possible during sowing or planting out, only a few more weeds may appear. Avoiding cultivation also helps to conserve soil moisture.

Chlorpropham/propham/diuron. This is a selective herbicide which acts through the soil and which is very similar to some of those often used in commercial vegetable production. It is applied evenly to the soil surface with a hand sprayer or through a fine rose after sowing, but before any weeds have appeared. It is taken up by the roots of the weed seedlings and it has a residual effect, remaining in the soil for several weeks and dealing with weeds that germinate during this period. It will not kill established plants, so the ground has to be free from weed vegetation to start with, and it needs moisture in order to work. In a dry spell, therefore, watering is advisable before application.

Chlorpropham/propham/diuron can be used after sowing a whole range of vegetable crops, as specified by the manufacturer's label. They include peas, beans, carrots, beet, brassicas and onions. For transplanted crops, application is made to the soil first and the plants then put out into the treated soil. On perennial crops such as asparagus and rhubarb, it can be applied in autumn after the foliage has died back and repeated in spring before growth starts again.

Propachlor. This is also a selective, soil-acting herbicide which kills annual weeds as they germinate over a period of six to eight weeks after application. It is available in the form of granules which are shaken lightly and evenly over the soil so as to achieve the recommended rate. This herbicide can be used on all the brassica crops and on onions, leeks, shallots and dwarf and runner beans. On sown crops it is best applied immediately after sowing; on transplanted crops, within two days after planting out. Propachlor does not kill weeds that are already established, and so the ground must be free from weed vegetation and also without clods. Like other soil-acting herbicides it needs moisture in order to work, although application under very wet and cold conditions should be avoided.

Propachlor controls most of the annual weeds that occur in quantity in vegetable gardens, including annual meadow grass, groundsel, shepherd's purse, chickweed and speedwells. Application can be repeated after six weeks or so in order to prolong the effect, first removing by hand any weeds that have survived the initial treatment.

Simazine. This is another soil-acting herbicide which is applied to the soil surface through a sprayer, dribble bar or fine rose and which persists for a period of months, killing weeds as they germinate. However, most vegetables are susceptible to simazine and there are only three for which it is recommended. For sweet corn it is applied within a week after sowing and firming the seedbed. The ground is then left undisturbed. It is advised that no other crop should follow within seven months. In established asparagus beds, simazine can be very useful for keeping down annual weeds throughout the growing season. To do this, the beds are cleaned up in spring, simazine applied according to directions before the spears begin to emerge, and the soil is then left undisturbed. Rhubarb usually controls weeds effectively by itself because of the dense leaf canopy which it provides. Simazine can be used if necessary, however, provided the crop has been established for at least a year. It can be applied in autumn or in early spring. Care is needed to ensure that the proprietary product contains only simazine; those sold for total weed control on paths and which contain other chemicals as well are not safe for use on crops.

PERENNIAL WEEDS

These are weeds which always persist for more than a year, and often do so for many years. The two unfavourable periods which all plants have to contend with are winter, with its sometimes severe cold, and summer with the possibility of drought. Annual weeds make sure of the continued survival of the species during extreme conditions by maintaining reserves of seeds in the soil which can give rise to new plants when conditions become favourable again. Those perennials which are not frost-hardy have underground storage organs, so that even though the above-ground shoots die down in winter, new ones can develop as soon as the soil warms up in spring. Many of them are also deep-rooted, and can draw water from below so that they are able to thrive during dry summers when other plants are at a disadvantage.

Some perennial weeds spread entirely by vegetative means and either produce no seeds at all, like slender speedwell, or do so only occasionally, like hedge bindweed. Others produce seeds in considerable numbers each year, and may rely on them for colonizing new areas. Perennial weeds in general are discouraged by the cultivations normally involved in growing vegetables, and many of them are not a problem in the vegetable garden. This is true of daisy and ribwort plantain, for example, which are common weeds of lawns. However, there are some exceptions to this rule and it is these which are a particular nuisance among vegetable crops.

From the practical point of view of controlling them and minimizing their effects there are three characteristics of perennial weeds which are especially relevant: first, whether they spread by vegetative means, and if so how; second, whether the perennating parts of the plant which persist from year to year are above the ground, in the surface soil, or deep down in the soil; and third, whether or not they produce seeds.

Weeds with tap-roots

These produce a large, simple or branched, vertical storage root, bearing one or more rosettes of leaves at its crown from which the flowering stems arise. Common examples are dandelion and the curled and broad-leaved docks. There is no natural vegetative spread, but in a garden, regeneration from roots cut up by cultivations can occur. Both produce numerous seeds which enable them to spread. Those of docks are very persistent in the soil, more so than those of dandelion, though even these can remain dormant for several years in cultivated soil. Another tap-rooted weed often found on the sites of old gardens or allotments is horse-radish. The roots have considerable powers of regeneration and this plant is difficult to control by cultivation; however, it does not usually produce any seeds in this country.

Creeping weeds

These include the most troublesome perennial weeds of gardens. They produce extensive systems of creeping stems or roots which enable them to spread rapidly and colonize large areas. In some weeds the creeping stems are in the form of stolons or runners above the ground. Plantlets are borne at intervals and these become rooted like strawberries and eventually become detached from the parent. In others, the creeping stems are below the ground and are called rhizomes. They bear small scale-like leaves with dormant buds, and these can be stimulated into growth by fragmentation of the rhizome during cultivation. Finally, there are the weeds in which there is an underground system of horizontal and vertical roots, which can penetrate to considerable depths. These can produce new buds at any position along their length and are often brittle, so that cultivation can easily result in new plants while the old ones regenerate from the deep roots.

Characteristics of some common creeping weeds

Creeping buttercup. Has above-ground creeping stems or runners which bear new plantlets. Some leaves remain green through winter. Also spreads by seeds.

Slender speedwell. Originally introduced as a rock-garden plant; has bright blue flowers. Now a common lawn weed and can spread into cropped ground. Stems spread on the soil surface and root at the leaf joints. Does not produce seeds.

Couch. The commonest perennial grass weed. Has branched

underground rhizomes with dormant buds usually only shallow, within a spade's depth. Shoots remain green in winter. Produces seeds, but main spread in gardens is by fragmentation of rhizomes during cultivation.

Black bent. Perennial grass like couch. Has branched underground rhizomes with dormant buds, usually only shallow. Shoots remain green in winter. Feathery flowering heads produce large numbers of very small seeds which readily germinate.

Stinging nettle. Has branched, yellow roots and horizontal rhizomes just below the soil surface. Short shoots remain green in winter. Produces quantities of seeds which can persist in the soil.

Ground elder. Very common garden weed. Has spreading system of underground rhizomes with dormant buds, usually shallow. Shoots die down in winter. Sometimes produces seeds, but main spread is vegetative by pieces of rhizome.

Colt's-foot. Has spreading system of underground rhizomes, often deep in the soil, with dormant buds and shoots bearing clusters of buds. Yellow flowers appear first in spring, followed by large leaves which die down in winter. Produces many seeds. These are dispersed by wind and can germinate immediately if they fall on moist soil, but they do not persist.

Creeping thistle. One of the commonest perennials. Has underground system of brittle, horizontal storage roots from which shoots arise; also deeply penetrating vertical roots. Shoots die down in winter. Patches of plants usually all male or all female; seeds sometimes produced, but not important as a means of spread in gardens.

Perennial sowthistle. Has underground system of brittle,

horizontal storage roots from which shoots arise; also deeply-penetrating vertical roots. Shoots die down in winter. Tall stems with yellow flowers produce seeds which are dispersed by wind.

Field bindweed. Very common and persistent weed. Has underground system of deep-penetrating roots from which clusters of shoots arise. Branched shoots initially lie flat on ground, then twine up crops. Produces many pink and white flowers, but seeds not usually important as a means of spread in gardens.

Hedge bindweed. Often invades from hedges or fences; very persistent. Has underground system of branched rhizomes with dormant buds, often deep in the soil. The twining shoots die down in winter. Flowers white, larger than those of field bindweed, but rarely produces seeds.

Field horsetail. One of the most troublesome and persistent weeds where it occurs. Has underground system of horizontal rhizomes, with vertical penetration deep in the soil. Also has tubers. Not a flowering plant; pinkish-brown fertile stems with spore-bearing cones appear in spring. Followed by erect, spiky green vegetative shoots which die down in winter.

Mechanical control

Tap-rooted weeds are comparatively easy to deal with. By inserting a garden fork at one side and levering, while at the same time grasping the crown of foliage and pulling, they can usually be removed completely. Docks do not regenerate even if the root breaks, provided that the upper 4 in. (10 cm) or so is removed. Even the thinner parts of dandelion roots, however, will form new buds if they are allowed to remain, so that it is important to get out as much of the root as possible. Horse-radish also will regenerate from any parts of the root left behind. Weeds like creeping buttercup in which the creeping stems are above ground are quite easy to get rid of, although it means that every plantlet must be dug out, and vigilance is needed to ensure that they do not establish again from the seeds which are probably present in the soil.

Those weeds which have underground systems of creeping rhizomes or roots are a much more difficult proposition. When the rhizomes are near to the surface, as they are in couch, digging out is feasible on ground used for growing annual crops. It is much better to use a fork than a spade for this purpose, since the latter tends to cut the rhizomes into short lengths. These are more difficult to remove, and any left behind will simply sprout again and soon form new plants. An alternative is to try to break up the rhizomes into pieces deliberately, by rotary cultivation for example. This causes a lot of the dormant buds to sprout and the process is then repeated several times as soon as new growth appears. The danger with this is that unless done very thoroughly, the problem can be made worse rather than better. Another alternative for couch is to bury the rhizomes, but it is essential to ensure that all of them are put down to a depth of at least 6 in. (15 cm).

Digging out is not a practical method of coping with those weeds which penetrate more deeply into the soil. Although the shallower roots and the vertical shoots which develop from them can be removed in this way, there are certain to be thick storage roots much further down than a spade's depth. So that while digging may bring about temporary relief, it is not a permanent solution, and it does of course require considerable time and effort.

In theory, hoeing conscientiously carried out and continued over a period of several years will weaken the plants and gradually eliminate them. Food reserves in the storage organs are used up in sending shoots to the surface and if these are hoed off before they have had time to reverse the process and replenish the reserves, then the weed ought to get weaker. Attacking a weed like creeping thistle just before it flowers, when the food reserves are at their lowest, might also seem sensible. In practice, however, hoeing is not a good way of getting rid of perennial weeds like this; it takes a long time, and some shoots are inevitably missed. It is a common complaint that even assiduous hoeing for many years has failed to make any real impression on an infestation of field bindweed or field horsetail.

There is one weed, oxalis, which is greatly encouraged by hoeing. Fortunately it is not very common, but is extremely serious where it does occur in gardens. The plant does not creep, but produces large numbers of small bulbs (bulbils) which are only loosely attached to the parent. Hoeing, or indeed any soil disturbance, only tends to dislodge them and cause new plants to appear over a wider area. The important thing is to recognize it as soon as it appears in the garden, to remove it carefully together with the soil round it, and then burn it.

Other physical methods of control which can be used against annual weeds are of no avail against those perennials with underground root or rhizome systems. Mulching is not effective, and this is not surprising when it is remembered that shoots of field bindweed can come through an asphalt pavement. Flame-weeding too, like hoeing, has only a temporary effect.

Chemical control

It is in the control of troublesome perennial weeds that chemicals can be especially valuable in the vegetable garden. To get at these weeds through the roots would require high rates of chemical which would be quite out of the question on ground needed for growing vegetables. Fortunately, however, there are herbicides which are described as being 'translocated', that is moved from place to place within the plant. When applied to the weed foliage, they are taken into the leaves and then move down the stem into the underground parts. Not only are the shoots which are actually treated ultimately killed, but an appreciable part of the below-ground rhizomes or roots will be affected and will be unable to send up more shoots in the future. This is the great advantage of these herbicides; they are actively transported by the plant with the food materials being manufactured in the leaves down to the parts that otherwise cannot be reached. It follows from this that there are two basic requirements for using these herbicides effectively. First, sufficient foliage must be present at the time of treatment to retain an adequate amount of the chemical. Second, the leaves must be manufacturing food and exporting it from the shoots down into the storage organs. So the golden rule is to treat the weeds when they have well-developed shoots with plenty of foliage and are actively growing.

Translocated herbicides for the control of perennial weeds are not selective in vegetables; if any solution gets onto the crop foliage damage will occur. There are two ways, therefore, in which they can be used. The first involves treating the weeds at a time when no crop is present in that particular area. If the plot is fairly heavily infested, early-maturing crops can be harvested and the weed vegetation left behind. This section of the garden can then be treated and after waiting long enough for the herbicide to take effect, late sowings or plantings can be made. Normal crop rotation will allow other sections to be treated in the following year, so that eventually control is achieved over the whole plot without interrupting the cropping schedule. If, on the other hand, there are only scattered shoots or patches of perennial weeds, spot treatment would be the appropriate technique. This involves treating the shoots individually, and if this is done with care, even shoots growing among the crops can be successfully dealt with. Run-off of liquid from the weeds must be avoided; apart from the risk of crop damage, any herbicide solution that falls on the soil is simply wasted.

As with the chemical control of annual weeds, the manufacturers' directions should always be closely followed, not only in relation to the strength of the solution and method of application but also for safe handling and storage of herbicides. The following paragraphs are intended to indicate the possibilities there are at present for chemical control of perennial weeds in the vegetable garden.

2,4-D. This is a component of various proprietary products sold for the selective control of lawn weeds. Some of these contain 2,4-D only, in others it is combined with other 'growth-regulator' or 'hormone' herbicides such as mecoprop or dicamba. In the vegetable garden they can be used only for spot-treating weed shoots. If a small amount of solution is made up at twice the strength recommended for lawns it can be applied to the weed shoots with a paint brush. Aerosol products which deliver the herbicide as a foam are also useful for spot treatment. 2,4-D is very effective against dandelion, creeping thistle and field bindweed, but it has no effect on grass weeds and only limited activity against some of the worst weeds like ground elder and field horsetail.

Dalapon. This herbicide is effective against grasses, and its particular virtue is in the control of couch. Dalapon can be used in established asparagus beds, applying either in spring before the spears emerge if the couch is present and growing actively, or as a spot treatment to the weed foliage in summer when the asparagus fern has fully developed. Contact with the fern should be avoided. It can also be used in rhubarb, applied in autumn or early spring when the crop is dormant but the grass is still growing, avoiding contact with the rhubarb crowns.

Glyphosate. This herbicide has been introduced only recently, and it has proved particularly useful to the gardener for two reasons. First, when used correctly and with a knowledge of the conditions under which it will perform best, it is a very effective chemical. Second, it is active against many of the problem weeds which up to now have been difficult to deal with. When used on a part of the garden which is not being cropped, it can be applied with a dribble bar, through a fine rose on a watering-can, or with a garden sprayer. Best results are usually obtained with a sprayer because more of the solution is retained on the weed foliage, but great care is needed to avoid any drift of spray droplets onto adjacent crops. As already emphasized, the weeds should have plenty of foliage and be growing vigorously. It is also important that the foliage should be dry at the time of application and that there is no rain for six hours in order to give time for the chemical to be absorbed into the plants. Glyphosate acts slowly, and it may be several weeks before the weeds look as though they are really dying. The ground should preferably not be cultivated for at least a week, so that the herbicide has time to be translocated downwards in the shoots to the roots or rhizomes. Any glyphosate which reaches the soil is inactivated almost immediately however, and if necessary crops can safely be sown or planted the day after application.

Some weeds, like couch, are normally killed completely by a single treatment. Others, such as field bindweed and field horsetail which have very extensive underground systems and relatively little foliage require more than one treatment. Spot treatment of individual weed shoots growing among crops is a useful method of following up a glyphosate treatment or preventing perennial weeds from becoming a more serious problem. The solution can be painted on, taking care to avoid all contact with the crop. Spot treatment has now become much easier with the advent of a brush-on gel formulation of glyphosate produced especially for the purpose.

LIVING WITH WEEDS

Weeds are survivors, well equipped to cope with life in a disturbed habitat and able to take advantage of any opportunity there is to establish themselves and multiply. There is no question of getting rid of all the weeds completely; it is a matter of control, of so organizing things that they do not get out of hand. There are two phases to this. The first is a short-term one of bringing a new or a very neglected garden into a state where it is possible to grow vegetables without serious difficulty. The second is the unending struggle to keep it that way.

New gardens

It is likely that there will be a mixed vegetation of various grass and broad-leaved perennial weeds, and that the soil will contain large numbers of living seeds of both annual and some perennial weeds. It is easier to make a determined attack on the perennials right at the start than it is later on once cropping has begun. How this is tackled depends partly on the kinds of weeds that are present, and partly on how urgent it is to start cropping. Possible approaches include:

Digging. Laborious; although annual weeds and couch can be usefully buried, not effective against deep-rooted perennials which will have to be dealt with later. Digging can be made easier by treating with paraquat/diquat to kill the vegetation which can then be skimmed off before starting.

Rotary cultivation. Quick, but tends to propagate perennial weeds by chopping up roots and rhizomes. Best if done several times at intervals of a few weeks. Not effective against deep-rooted perennials.

Treatment with dalapon. Useful when it is mainly grass weeds that are present. Sowing or planting can take place after six to eight weeks.

Treatment with glyphosate. Very effective way of dealing with both grassy and broad-leaved perennials; also kills annual weeds. Best to apply when perennials are well-grown and approaching the bud stage. Sowing or planting preferably should be delayed for three to four weeks to allow the herbicide to exert its full effect, but can be done sooner.

Treatment with ammonium sulphamate. Best applied to well-grown vegetation during a dry spell. Corrosive to metals, so plastic sprayer or watering-can needed. Kills a wide range of annual and perennial weeds, but not some deep-rooted ones like field bindweed. Sowing or planting can begin after eight weeks, or twelve weeks if it stays very dry.

Treatment with sodium chlorate. Effective total herbicide, killing all kinds of weeds. Best to use a product with a fire depressant in it. Vegetation, wood and clothing sprayed with sodium chlorate become highly inflammable when dry, so great care is needed in handling and application. Sowing or planting has to be delayed for up to twelve months after treatment.

Whatever approach is adopted, there will probably be some perennials which escape or are not completely killed, so that in the following years a campaign of spot treatment is necessary.

Established gardens

Since the development of translocated herbicides, and of glyphosate in particular, it has become much easier to contain perennial weeds than ever before. By diligent spot treatment, which is after all a very easy operation, it should be possible to eradicate these weeds. There will, of course, be new introductions; creeping weeds will invade from hedges or the base of fences, or from the gardens of less horticulturally-inclined neighbours. Couch invading in this way is easy to deal with by making a vertical cut with a spade and then forking out the rhizomes. Shoots of deep-rooting perennials can be spot-treated with 2,4-D or glyphosate. Seeds of some common weeds, particularly dandelion and willowherbs, can be carried on the wind from some distance away so that there is a need to recognize perennials developing from seeds and to deal with them before they become established. Hoeing is only effective at the seedling stage; once beyond this, the young plants need careful forking-out. Asparagus beds offer a very favourable habitat for perennial weeds to establish from seeds brought in by the wind or dispersed by birds. If they are allowed to flourish, the useful life of the beds will be shortened, so that they need special attention.

Normal vegetable-growing practices are unlikely to introduce either seeds or root fragments of perennial weeds from elsewhere, although farmyard manure which has been stacked may well contain seeds of nettles. It is very easy, however, to introduce couch and ground elder into a garden in clumps of herbaceous plants, and once established, these weeds can readily spread into the vegetable plot. Introduction of the bulbous oxalis in this way is particularly to be avoided.

By judicious use of herbicides and a degree of vigilance it is possible to eradicate perennial weeds or at least ensure that they do not interfere with cropping. Unfortunately this is not true in respect of annual weeds. Prevention of seeding is the aim, and if this is achieved there is no doubt that the seed bank in the soil can be reduced to a low level. In practice, though, it is very difficult to get it down below a level of about 50 living seeds per square ft. (540 per square metre). It only takes a few weed plants to escape and reach maturity for the natural losses to be replaced; a few more, and there may be ten times as many seeds in the soil at the end of the season as there were at the start.

Most of the weed seeds to be found in the soil are produced by plants which have seeded on the spot. In a new garden there may be all kinds of weeds, depending to a large extent on what type of soil it is, what the land has previously been used for, and whereabouts in the country it is situated. The weeds growing on a light, well-drained sandy soil are likely to be very different from those of heavy clay soils, and there are differences in the kinds of plants to be expected in the north and west, for example, compared with the south-east. However, a survey of seeds present in the soil of fields which had been cropped commercially with vegetable for many years showed that there was a similarity in the weeds irrespective of location and soil type. Chickweed, annual meadow grass, annual nettle, groundsel, shepherd's purse, fat-hen and field speedwell were among the weeds most often recorded, and together these seven species accounted for 70 per cent of all the seeds found. As vegetable cropping proceeds, therefore, the character of the weed vegetation will change, and it is these species which are likely to feature increasingly as the main annual weeds.

Seeds of annual weeds can, of course, be introduced into gardens in various ways. Groundsel and the annual sowthistles have seeds which are readily carried by the wind, while those of black nightshade are carried from place to place by birds which feed on the berries. Manure stacks on farms often become clothed with annual weeds, and the traditional well-rotted manure may contain many seeds of plants such as red goosefoot. The little cruciferous weed hairy bittercress has explosive seed pods which scatter the ripe seeds for some distance, and in recent years this species has entered many a garden with containerized ornamental plants. The actual numbers of seeds brought in by these various agencies are likely to be very small in relation to the considerable quantities present in the soil. However, if they happen to be of species not already found in the garden they could, when once established, add to the existing weed problem.

Accepting that annual weeds are always going to be a nuisance to some extent, the question is how to keep them under control most easily. The one word which goes furthest towards answering this is 'timeliness'. Jobs done at the right time when conditions are most favourable give the maximum benefit with least effort. No matter whether it is hoeing, digging or the use of a herbicide, there is a best time in the particular circumstances; timeliness involves recognizing it and making the appropriate response.

Besides this, it is important to try to avoid making more difficulties than are absolutely necessary. If crops are sown in straight rows at a distance apart which is suitable for hoeing this helps to cut down the time needed for hand-weeding. Crop spatial arrangements which will obviously need a lot of hand-weeding are better avoided unless there is a suitable soil-applied herbicide available or the amount of hand-weeding is considered acceptable. The technique of covering the seed drill with compost (see p. 165) to improve crop emergence has the advantage of reducing the need for hand-weeding.

Virtues of weeds

In this chapter the main argument has been that weeds are, for various reasons, undesirable in the vegetable garden and that they should either be eradicated or at least kept down to the lowest feasible level. Perhaps it is appropriate to conclude by considering briefly some of the virtues of weeds, about which whole books have been written.

Our climate is such that bare ground which arises through natural processes like landslip or erosion quickly becomes colonized by plants. These plants cover the surface, their roots bind it; they extract mineral nutrients from the rocks and fix carbon from the carbon dioxide of the atmosphere; and when they die, their organic and mineral content passes onto and into the soil and enriches it. In this way fertility is built up. Weeds in a garden have the same effect; the difference is that here it is crops that we are trying to grow. Even in a vegetable garden, however, weeds are not necessarily 'plants out of place' all the time, and it seems sensible that we should take what advantage we can of the benefits they can confer.

On ground which is temporarily free from crops, the presence of weeds may be doing no harm so long as something is done about them before they reach the stage of seeding. So they could be left as long as possible before either being dug in, if digging is needed for other reasons, or killed with paraquat/diquat. It is true that their nutrient value may be only small in relation to what is needed to grow a crop but there is some benefit to the soil. Weeds which are removed from among crops can be composted so that both the mineral and organic contents are re-cycled. Even in a compost heap which heats up properly not all weed seeds will be killed, but in any case the weeds should be removed before they begin to produce seeds. It is usually advised that perennial weeds should not be added to the compost heap but should be burned instead. This seems a pity; even the rhizomes of couch and roots of thistles will rot down to a large extent, and any living parts that remain can easily be picked out and put on the next heap when the ripe compost is being dug out.

5 Managing your soil

Vegetables can be grown on a wide range of soils, but to obtain first-class crops careful attention must be given to the fertility and physical condition of the soil. Most gardeners are aware that fertility can be adjusted by appropriate applications of inorganic fertilizers or organic manures (see *Know and Grow Vegetables*) but do not always appreciate the basic importance of soil structure. It is the purpose of this chapter to provide a better understanding of the physical conditions required for successful crop growth and to discuss how they might be obtained by suitable soil management.

The knowledgeable gardener will be quick to point out that plants can be grown without soil by using peat-filled growing bags, various granular materials or even by bathing the roots in a continuous flow of nutrient solution. While these modern techniques can eliminate many of the structural problems associated with soils, they are expensive and costly on anything but a small scale, and do not eliminate the need for careful attention to nutritional and environmental requirements. For the extensive production of vegetables, the convenience of soil will ensure its continued importance in the foreseeable future.

What then does a plant require from soil in addition to the obvious needs of support and anchorage? Soil must contain the nutrients required for crop growth, act as a reservoir for water and allow the free passage of air. To utilize these reserves the plant needs a healthy, vigorous root system and this is only attained if the soil is free of pests and diseases and is easily penetrated by young growing roots. The importance of an actively growing root system cannot be over-emphasized. Roots are important not only as a means of absorbing water and nutrients but also for the manufacture of some of the growth-regulating hormones essential for the proper development of the plant.

Before considering how to manage the soil, we should take stock of our raw material and see what it consists of and how it varies from soil to soil.

THE NATURE OF SOIL

Basically, soil consists of a mixture of mineral particles varying widely in size but which are arbitrarily divided into three size ranges; sand, silt and clay in order of decreasing size. The term *texture* is used to describe the different proportions of these particles and should not be confused with *structure* which refers to the way in which the individual particles clump together. With practice, soils of different texture can be recognized by the feel of a moist sample kneaded between the fingers. Sand particles are gritty, silt is smooth and silky, while clay is sticky. If no one size of particle predominates then the soil is a loam. The proportions of sand, silt and clay in some representative soils are shown in Table 5.1. Gardeners often describe sand and silt soils as being 'light', meaning they are loose and easy to cultivate, whereas clay soils, being sticky, are called 'heavy'.

Soil particles are irregular in shape and do not pack tightly together but are separated by spaces or 'pores'. Surprisingly, these occupy between 40–60 per cent of the soil volume. If the soil is mainly sand, the pores will be large and free-draining,

Table 5.1 Approximate percentage composition of some representative soils

Soil type	*Sand*	*Silt*	*Clay*
Sandy loam	65	25	10
Loam	40	40	20
Silt loam	20	65	15
Silt clay loam	10	55	35
Clay	20	20	60

but if it is mainly clay then they are small and slow to drain. Most soils, however, consist of particles of a range of different sizes, and consequently the pores vary from small to large. Furthermore, soils do not usually consist of individual particles but are bound together to form aggregates or 'crumbs'; in other words they are structured. Thus, instead of having only small pores, a fine-textured soil will have small pores within the crumbs and large pores between them. This network of variable-sized pores is the basis of a well-structured soil and is most important to plant growth. The large pores empty quickly after rain and allow air to enter, while the small ones retain water which may later be taken up by plants. The large pores are also important for root penetration. Research has shown that roots do not readily enter pores unless the pores are of the same or larger diameter than themselves. Crops with fat roots like peas and beans are therefore more susceptible to soil compaction than those like cabbage and turnips with thin roots.

Soil is not simply a collection of mineral particles but also contains between 1 and 5 per cent organic matter (over 25 per cent for peaty soils). This is composed of the dead and decaying remains of plants and animals and is the food source of innumerable living micro-organisms. As fresh organic matter is decomposed by bacterial and fungal attack there is a gradual release of plant nutrients until finally a residue of dark-coloured material remains, called humus. Humus has an important role to play in soil structure since it binds the soil particles together and helps to stabilize the aggregates.

HOW CAN POOR SOIL STRUCTURE BE RECOGNIZED AND WHAT ARE ITS CAUSES?

As we have seen, good soil structure is of great importance to crop growth. Before considering how to maintain or improve structure, however, it is essential to assess the physical nature of your own soil. This can best be achieved by direct examination of the soil profile since not all defects are visible on the surface of the soil.

Dig a hole or trench to a depth of 18 in. (46 cm), if soil depth will permit, and examine the exposed face. If a network of visible pores and cracks is apparent throughout the profile, all is well since air and roots will be able to penetrate easily and excess water will readily drain away. If, on the contrary, the soil is tightly packed together and plant roots are confined to a few cracks or earthworm channels instead of appearing throughout the mass of the soil, then plant growth will almost certainly be restricted.

Compact soil can often be detected as a resistance when digging or may be revealed by a layer of wet soil occuring above the level of compaction because of reduced drainage. Compact soil is frequently found as a distinct layer or 'hard pan' at the boundary of the topsoil and subsoil. In severe cases, roots will be confined to the topsoil and will be seen spreading horizontally above the hard pan. Taproots of carrots and parsnips will be stunted, or kinked (Fig. 5.1).

Poor structure is often a consequence of soil texture. Light

Well-grown roots from good structured soil

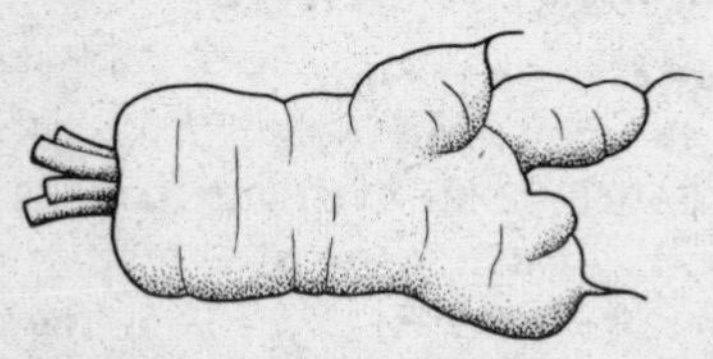

Stunted roots from compact or waterlogged soil

Fig. 5.1. Carrot roots (Amsterdam Forcing type) affected by soil compaction at 3 in. (7.5 cm) depth. A typical result of preparing a seedbed when the soil is too wet.

sandy or silt soils with low clay or organic matter contents have nothing to bind them together. Heavy rain will cause them to settle or 'slump' to a density which will restrict root growth. Frequently slumping is confined to the soil surface. Rain falling on a fine seedbed washes small particles from the soil aggregates into the pores between them. On drying, this surface will set hard to produce a crust or 'cap' which is sufficient to restrict or even completely prevent seedling emergence.

Hard pans occur naturally on some soils, particularly old heathland soils, where fine soil particles and iron compounds wash down the profile and are deposited at a depth of about 12 in. (30 cm) to form a hard impenetrable layer. More frequently, however, hard pans are man-made, as a result of working or travelling on wet soil with heavy machinery. This is a major problem for commercial vegetable growers who, because of market dictates, often have to cultivate and sow or harvest crops when soil conditions are less than ideal. Fortunately, severe hard pans are unlikely to arise in the garden although they may survive from previous agricultural use of the land or have been created during house-building. It is important to bear in mind, however, that soil structure is very weak when the soil is wet and compaction can result simply from walking on or trying to cultivate wet soil.

It was noted earlier that the organic matter content of a soil generally falls between 1 and 5 per cent. The exact content depends on a balance between the quantities of fresh organic material added to the soil and the losses due to decomposition. Some crops produce far greater residues than others. In particular, grass returns much larger amounts of organic matter to the soil in the form of root residues than most other crops. The organic matter content of a soil under grass is therefore high and soil structure is generally good. If the soil is then cropped with vegetables, the organic matter content will decline rapidly over the first three or four years, then level off until a new equilibrium is reached. This rapid decline will be due mainly to the lower return of root residues and to the removal of crops. It is often assumed that regular cultivation, by aerating the soil thereby encouraging microbial activity, would also increase the rate of organic matter decomposition but this has not been borne out by experiment. Whether a reduction in organic matter will be detrimental to soil structure depends to a large extent on the type of soil. Chalk and limestone soils are naturally well-structured and are not sensitive to organic matter levels. Other soils, particularly sands and silts, will become increasingly susceptible to slumping while clays will become more difficult to cultivate and will tend to dry into hard clods.

There is no practical way for the gardener to measure the organic matter content of his soil. In fact, even if it were possible, soil scientists would not be able to specify a minimum desirable level because of the considerable variation in effect from soil to soil. Instead, the prudent gardener must always be on the look-out for tell-tale signs of deterioration in structure. Are seedbeds becoming difficult to prepare? Does the soil no longer drain quickly after rain? Do crops wilt readily in short dry spells? Are yields poor despite correct feeding and the absence of pests and diseases? These are the danger signals. Early remedial action or a change in soil management may prevent later crop failure.

HOW CAN TEXTURE AND STRUCTURE BE IMPROVED?

Drainage

Good drainage is vital to successful soil management. We have already seen that wet soils are more liable to structural damage, but waterlogging also has a direct effect on plant growth. A soil which becomes completely saturated, so that even the large pores and cracks are full of water, contains no free oxygen. Roots will be unable to 'breathe' and are killed or seriously damaged and become more susceptible to attack by soil pests and diseases.

Improvements to drainage depend very much on the cause

of the problem (Fig. 5.2). Where soil inspection has revealed a hard pan or a layer of unrotted organic matter, a deeper cultivation than normal is all that is required. If the trouble is compaction in the topsoil the remedy is thorough loosening by digging or forking, followed by the regular addition of organic manures to help keep the soil open.

The most difficult drainage problem to overcome is an impermeable clay subsoil. Regular additions of organic materials mixed well into the subsoil will gradually increase the depth of workable soil. The farmers' solution of laying drains, although ideal, requires considerable skill in achieving the correct spacing and gradients, and the surplus water must be disposed of in a ditch or soak-away pit. In the small garden, a simple trench 12 in. (30 cm) wide and 24—36 in. (61—91 cm) deep, filled with rubble and dug at the lowest point of the site may be adequate.

Where drainage problems are infrequent, or in high rainfall areas, an alternative approach is to increase the depth of topsoil by creating a 'raised bed'. The vegetable area is divided into strips or 'beds' separated by pathways. Topsoil, removed from the paths, or imported, is added to the growing area to form an increased depth of free-draining soil. By keeping the beds to about 4–5 ft. (122–152 cm) wide it is possible to sow and harvest by reaching in from the pathways. In this way the cropped areas are never compacted by trampling. A similar system is used by commercial growers but in their case the pathways are wheel tracks, the beds being spanned by a tractor.

A wet soil is a cold soil and is slow to warm up in the spring. By improving drainage or by using raised beds it will be possible to sow many crops earlier in the year. Additional benefits may be obtained by placing cloches or other forms of protection over the area to be sown a few days before it is required. Also, the use of transplanted crops in the spring permits greater flexibility in timing since their use eliminates the need to prepare a fine seedbed when the soil is wet and easily damaged by being worked.

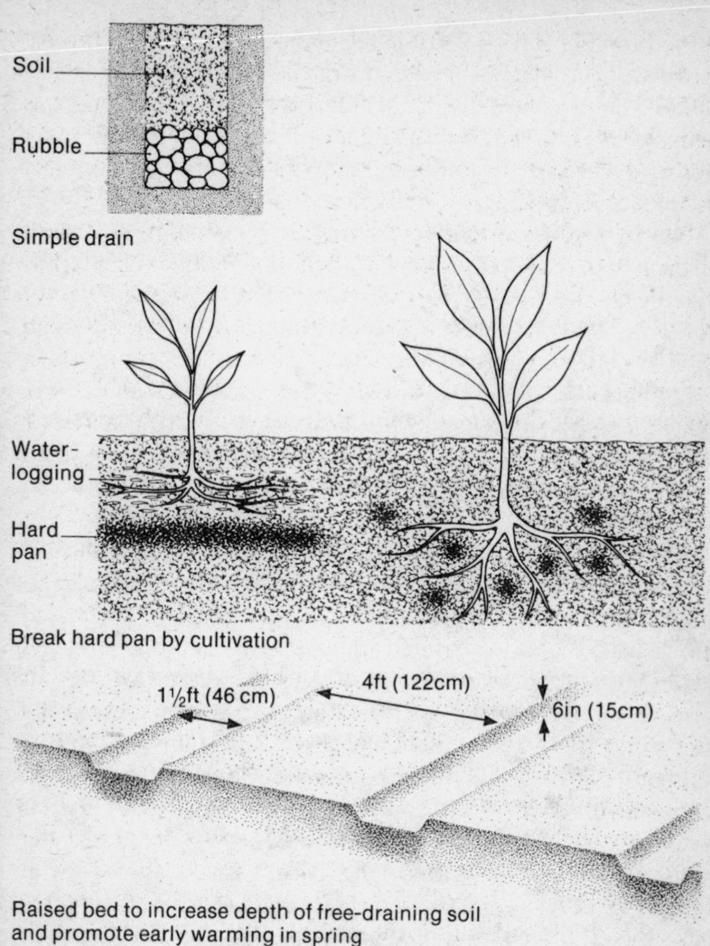

Fig. 5.2. Improving soil drainage.

Altering texture

Gardeners often wonder if it is possible to improve a sandy soil by adding clay. Many years ago this was a common agricultural practice and was referred to as 'marling'. It is now believed that the benefits of marling were due mainly to the lime content of the clays used rather than to any marked change in soil texture. Some simple calculations will show the impracticality of altering texture in this way.

Suppose you have a sandy-loam soil containing 67 per cent sand, 23 per cent silt and 10 per cent clay. The soil in a plot 9 × 3 yards (8.2 × 2.7 metres) to a depth of 12 in. (30 cm) would weigh about 10 tons (or tonnes). An addition of 1 ton of clay, if it could be thoroughly mixed in, would raise the clay percentage to 18 per cent. The soil would still be predominately sandy! Similar calculations may be made for the addition of sand to clay soils.

Small improvements in the water retention of coarse-textured soils can be made by adding ashes, but this practice is not recommended since many ashes, particularly those of industrial origin, may contain toxic materials.

If it is accepted that texture cannot easily be altered, we must learn to manage the soil we have and to channel our efforts into improving its structure.

Organic manures

Many gardeners rely on bulky organic manures such as compost or farmyard manure to maintain or increase the organic matter content of their soils and thereby preserve or improve structure and workability. Research has shown, however, that the amounts generally applied are often insufficient to have the desired effects. To increase the level of organic matter in a soil requires the *annual* addition of manure at a rate of at least 10 lb. per square yard (5.4 kg. per square metre). As soon as you stop applying manure the organic matter level will again start falling towards a new equilibrium. We must ask ourselves, is it practical to apply such large amounts regularly and are the benefits worthwhile?

Studies which have attempted to answer these questions are complicated by the fact that most organic manures contain appreciable amounts of plant nutrients. To test whether they have an advantage over inorganic fertilizers requires careful balancing of the nutrients applied. This is easier said than done, but the results of carefully-conducted experiments show that yields on some organically-treated soils are slightly higher than on those soils receiving inorganic fertilizers alone. To some extent this is due to differences in the distribution and availability of nutrients in the soil; manures generally are ploughed or dug in to a greater depth than fertilizers, which are applied during seedbed preparation. Some of the additional benefit, however, is attributable to improvements in the physical condition of the soil. In particular, regular, heavy applications of well-rotted manures will increase water retention of coarse-textured soils, while more fibrous materials will help to keep clay soils open and improve their workability. To achieve these effects, it is preferable for the manure to be mixed throughout the topsoil. Applying it as a layer at the bottom of a trench, as is often suggested, should be avoided since this can reduce aeration, cause waterlogging, and restrict root growth, and does not maximize the effect on physical conditions.

If supplies of organic manures are limited, do not spread them thinly but apply them at high rates to those crops which are known to be most responsive such as leeks and runner beans. This is one consideration in planning crop rotations and is examined towards the end of the chapter.

Farmyard manure. Farmyard manure, or FYM as it is often referred to, is perhaps the most widely available soil amendment but can be expensive to obtain in urban areas. It consists of animal dung and urine, usually from cattle, together with straw or other animal bedding, which has been stored and rotted. Its value to the soil is largely due to its nutrient content but this can vary widely depending on how the manure has been made and stored. Nitrogen and potassium, and to a lesser extent phosphate, are washed from the manure heap if it is exposed to rain, and much of the original nitrogen can be released to the air as ammonia or nitrogen gases during decomposition. Obviously, the gardener has no control over the production of FYM but its variable nature emphasizes the need for occasional soil analysis using a proprietary soil test kit so that any nutrient deficiency can be made up with additional fertilizer.

To reduce the loss of nutrients caused by winter rainfall, well-rotted FYM is best incorporated in the soil in late winter or early spring. Manure with a high proportion of unrotted straw can cause a temporary deficiency of nitrogen as it breaks down because of a rapid increase in micro-organism activity and the 'locking-up' of nitrogen as microbial protein. Strawy manure is best, therefore, applied in late autumn. If incorporation just before sowing time is unavoidable, extra fertilizer nitrogen should be used to help overcome any temporary nitrogen deficiency.

FYM, together with most other organic manures, should not be applied at the same time as lime because chemical reactions will waste nitrogen which is released as ammonia gas.

Garden compost. Well-made compost is a good substitute for FYM and can contain slightly more plant nutrients. In some gardening books compost-making is treated almost as an art since although the scientific principles involved are well known it is often difficult to provide the conditions necessary to produce a good humus-like material.

Composting involves the partial decomposition of vegetable matter by micro-organisms and requires air, moisture and heat. How these conditions can be met are dealt with in great detail in the gardening literature and will not be repeated here. However, it is appropriate to consider some of the problems which arise in attempting to produce good compost on a scale suited to the small garden. Temperatures of about 150°F (66°C) are necessary throughout the compost heap in order to ensure rapid decomposition of the material and to kill unwanted weed seeds, pests, and fungal spores. Inevitably, the temperature of a heap will be greatest in the centre, reducing towards the sides and top. This problem will be much worse with a small heap because of the high ratio of surface area to volume. Enclosure in a purpose-made compost bin to retain heat will help overcome this problem, but ideally it should be emptied occasionally and the heap turned in order to introduce air. In situations where compostable material is only available in small quantities and over a period of time, high temperatures are rarely attained and the compost will inevitably be very variable in quality. In conclusion, the main difficulty in using compost is in producing it in the large amounts needed to give a worthwhile improvement in soil structure.

Peat. Peat is a good source of organic matter and is particularly effective at opening up clay soils, where its effects last longer than those obtained with FYM. However, it is low in plant nutrients, and when used at high rates can make the soil acid. It is also expensive and hence is not usually practicable unless a cheap local supply is available.

Other organic manures. A number of alternative organic manures may be available to the gardener. Their value should be judged on cost and on their content of plant nutrients but some of them have drawbacks and require care in their use.

Straw and sawdust are both readily available but low in nutrients and can cause temporary nitrogen deficiency unless composted before being added to the soil. Sawdust from timber treated with wood preservatives should be avoided.

Sewage sludge and municipal compost are available in some areas and can be cheap. However, some samples, especially from industrial areas, contain high levels of toxic metals which can accumulate in acid soils and could be taken up by plants if regular, heavy dressings are used. Permissible levels of metal contaminants are controlled by regional codes of practice and anyone considering using these materials should contact their local Water or Municipal Authority for advice.

Poultry manure is often available cheaply but can be smelly and unpleasant to handle. It has the advantage of being rich in nitrogen and phosphate but unless mixed with straw or sawdust will have no effect on soil structure. Fresh droppings may release ammonia in amounts harmful to young plants, unless composted or added to the soil at least a month before sowing.

Spent mushroom compost is a well-rotted manure with about the same nutrient content as FYM. However, some samples contain lime and these should not be used regularly on neutral or alkaline soils since excess lime can cause deficiency of some minor elements.

Organic fertilizers such as hoof and horn, meat and bone meal, and dried blood are used only for their nutrient value and have no value in improving soil structure.

Green manuring. Many gardening books recommend green manuring as an alternative to a grass break. Crops such as mustard, rape or vetch are grown and dug into the soil when about 10 in. (25 cm) high. However, it is difficult to increase the organic matter content of the soil in this way unless green manure crops are taken in successive years.

Any crop grown for digging-in must establish quickly after sowing and grow rapidly in the remaining season, possibly being fitted in after lifting early potatoes. Care should be taken to prevent carry-over of soil-borne pests and diseases. Rape and mustard belong to the same plant family as Brussels sprouts, cabbage and cauliflower and should not be grown on ground infected with clubroot.

Experiments suggest that most of the benefits obtained by green manuring are through its effects on nutrient supply. The green manure crop will take-up soil nutrients, particularly nitrogen, which might otherwise be lost in the drainage water between harvesting one crop and sowing the next. Once dug in, the young leafy crop will rapidly decompose and release nutrients into the soil. To take advantage of this, the next crop must be sown soon afterwards. In practice, however, the fast-growing green manure crop will have been using water from the soil and this may result in difficulty in establishing the next crop.

It is often claimed that deep-rooted green manure crops such as comfrey benefit succeeding crops by making available nutrients from the subsoil. This seems unlikely. Most vegetables have root systems which penetrate much deeper into the soil than gardeners generally realize. Research has shown that the roots of many vegetables growing in a free-draining soil will reach a depth of at least 4 ft. (1.2 metres) by the time the crop is harvested.

Many of the benefits of green manuring may be more easily obtained by applying fertilizers. The root residues of a well-grown vegetable crop will to a large extent help to maintain structure. Beneath each square yard of a turnip crop is 12½ miles (24 km per square metre) of root!

Liming and soil structure

Lime is essential to soil fertility both to provide a plant nutrient (calcium) and for correcting acidity, and these aspects have been fully discussed in the companion volume to this book (*Know and Grow Vegetables*). It is widely believed that liming is also beneficial in stabilizing the structure of non-calcareous clay soils and reducing their stickiness, but the experimental evidence for this is not conclusive. Nevertheless, liming, by promoting crop growth and encouraging earthworm and micro-organism activity, will have indirect effects on soil structure. First, through the binding action of a vigorous root system and second, by improved breakdown of organic matter to humus.

Old gardening books often recommended gypsum (calcium sulphate) as a soil conditioner but apart from its well-known use in reclaiming soils which have been flooded by sea water, its effects on normal soils are likely to be negligible.

CULTIVATIONS

Cultivation is perhaps the most effective and immediate way of producing soil conditions favourable to crop establishment and growth. Yet the trend in agriculture over the last decade has been towards a reduction in cultivations. This has prompted many gardeners to question whether the cultivations traditionally associated with the vegetable garden, particularly deep digging and regular hoeing, are entirely necessary. Before answering this question, we will consider why cultivations are normally carried out and examine the background to this recent trend.

Traditionally cultivations have served two main purposes, to remove weeds and to produce a suitable tilth for plant growth. Obtaining a tilth invariably involves two stages, an initial ploughing, or digging, aimed at inverting the topsoil to bury weeds, followed by a number of secondary cultivations to create a seedbed. In recent years the development of effective herbicides has given the grower the option of controlling weeds without resorting to the plough or spade. This allows a reduction or even elimination of secondary cultivations with a greater reliance placed on the natural processes of structure formation. Unfortunately, systems of reduced cultivation are not without problems. Many soils, particularly sands and silts with a low organic matter content, gradually settle and become compacted when uncultivated to such an extent that root growth may be restricted. This is not so critical for crops with a long growing season such as winter cereals, but for short-season vegetable crops rapid root extension, as we shall see later, can be important. Nevertheless, experiments have shown that many vegetables can be grown successfully with reduced cultivations on well-drained soils, providing that care is taken not to damage soil structure at harvest since there are no opportunities for remedial cultivations. Although this is an important consideration for the commercial grower, in the garden, with hand-harvesting, the problem should not arise. However, the gardener is seldom likely to benefit from the main advantage of reduced cultivation systems, which is speed of operation and which enables the farmer to sow large areas rapidly whenever soil conditions are ideal.

Before considering whether reduced cultivations have a part to play in managing soil in the garden, let us examine the various cultivation methods and cultural techniques available to the gardener.

Single digging

Digging is the main cultivation in the garden and aims to invert the topsoil so as to provide a surface free of weeds and crop debris ready for the preparation of a seedbed. Digging also breaks up the soil to produce an open structure through which water and air can enter and in which roots can grow and proliferate. It is also used as a means of incorporating lime and bulky manures.

Single digging involves working the soil to a depth of about 10 in. (25 cm) using either a spade or a fork. The choice is one of personal preference but a fork is often easier to use on heavy or stony soils. As a general guide soil should only be worked if it is dry enough not to stick to your boots. This means that heavy soils are best dug in early autumn, before they have been wetted by winter rains.

The soil should not be worked initially to a fine tilth since it will recompact during the winter and lie wet in the spring causing delays in early sowings or plantings. Instead, it should be roughly levelled and left as fist-sized clods. This will allow rain to penetrate and the clods will gradually weather and break down under the action of alternate freezing and thawing to form a 'frost-mould'. If larger clods are left, there is a danger that when the soil is worked in the spring the frost-mould will disappear into the cracks and voids. This would create a patchwork of weathered and unweathered soil which is difficult to work down to an even seedbed. If digging is unavoidably delayed until the spring, clods should be broken down as digging proceeds.

Some gardeners may possess or be contemplating buying a mechanical cultivator. They are most useful for seedbed preparation rather than as a substitute for digging since they tend to pulverize the soil and create a fine tilth. If they are used regularly for all cultivations there is a danger of the tines creating a hard pan at the normal depth of working.

Double digging

This consists of working the soil to a depth of about 20 in. (51 cm) and involves first opening a trench to the depth of a spade and then forking the exposed subsoil, care being taken to avoid mixing top- and subsoil. This laborious process was a matter of routine in the vegetable garden at the turn of the century but today's gardener may ask if the effort and time involved are worthwhile, in view of the trend towards reduced cultivations. But it should be realized that high yields will only be obtained consistently if root growth is not impeded by subsoil compaction. Therefore, if soil inspection has revealed a hard pan, deep cultivation is essential.

Recent research has shown benefits of deep cultivation on well-drained soils which are not visually compact and in which roots normally penetrate to at least 36 in. (90 cm). These benefits are sometimes spectacular. In one experiment the yield of broad bean pods was increased by 95 per cent. While this was exceptional, yield of a range of vegetables has generally been increased by 10 to 30 per cent. Records show that these results are due mainly to improvements in water supply. Plants take up water from the soil through their roots. By loosening the soil we enable the roots to penetrate deeper more rapidly. In the event of a dry spell, the water reserves in the subsoil are immediately available for uptake. This means that the effects of double digging will be small in wet years or if regular watering is practised, since in those circumstances there will be no shortage of water in the top-soil. Double digging can therefore be regarded as an insurance against drought and the ban then imposed on the use of garden hosepipes.

Double digging need not be carried out annually, for although the soil gradually resettles, the effects on yield in one series of experiments persisted for at least four years. Stabilizing the loosened subsoil by incorporating bulky organic manures would probably extend this period, and would also increase its fertility. This is particularly valuable in dry weather since plant roots can only absorb nutrients which are in solution in the soil water. As the topsoil dries out, the nutrients it contains become progressively less available. However, if there are roots and nutrients at depth then uptake can continue from the moist subsoil.

'No-digging'

The 'no-digging' system, strongly advocated by some gardeners, is similar to the reduced cultivation technique of the farmer in that it relies on natural processes of structure formation. It differs in that earthworms are actively encouraged by annually covering the soil surface with a layer or 'mulch' of organic manure to a depth of 1–2 in. (2.5–5.0 cm) Under these conditions the earthworm population flourishes and by dragging the organic material into their burrows they gradually build up the fertility and structure of the topsoil without the necessity of digging. The earthworms which are most useful for this task are the large burrowing species and not the small, red manure-worms or brandlings found in the compost heap. Gardeners should be cautious of so called 'commercial' earthworms sometimes offered for sale as soil improvers. They are invariably surface-dwelling species which live in high organic matter situations and do not survive for long in normal soil.

Providing that the large amounts of organic manure required – 1½ cubic feet per square yard (0.05 m^3 per square metre) – are obtainable, mulching as a technique has much to commend it. By protecting the soil surface from the drying effects of sun and wind, a mulch will reduce the amount of water lost from the soil by evaporation. Consequently, the soil beneath the mulch will remain moist, stimulating micro-organism activity and improving nutrient uptake. A mulch will also protect the soil surface from the destructive effects of heavy rain, and have an insulating effect which will keep the soil cooler during the day and warmer at night. A further benefit, providing the soil is free of perennial weeds and the mulch material is weed-seed free, is that mulching will help suppress weed growth and will thus reduce the need for hoeing.

Many of the benefits obtainable by mulching with organic materials can also be achieved by using a clear or black polythene mulch but, of course, the latter do not contribute to soil structure or fertility and may not be aesthetically desirable in the garden.

'No-dig' systems, whether based on mulching or simply on reduced cultivations are unlikely to be successful on all soils unless structural defects can first be remedied by careful cultivations and by incorporating bulky manures. Those soils where the surface layers are weakly structured and prone to slumping, or those with poorly-drained subsoils are those least likely to be suitable.

Seedbed preparation

The preparation of a seedbed is the most important cultivation carried out in the vegetable garden. Failure to produce a 'good' seedbed will result in poor seed germination or seedling establishment and, inevitably, lower yields. In spite of its obvious importance, little research has been directed towards defining the ideal tilth for each crop since in practice it would be difficult to produce a specified tilth given the wide range of soil and weather conditions encountered. Nevertheless, we know enough of the requirements for germination and establishment for us to give reliable guidelines to the soil conditions desirable. Let us start by considering these requirements and then how they might be met.

For a seed to germinate it requires warmth, air and water. A dry seed placed in contact with moist soil will immediately begin to absorb water. If the seed is viable, air is present, and the soil sufficiently warm, then the seed's food reserves are mobilized and growth begins. Germination culminates in the outer seed-coat rupturing as the young root thrusts downwards and the shoot upwards. The young seedling is now at a critical stage. Rapid root growth is essential if the uptake of water and nutrients is to be adequate once the seed reserves are exhausted. Already we can appreciate what is required from the seedbed. First, the seed must be in good contact with the moisture in the soil if the initial water uptake is to be rapid. This is easier to achieve if the tilth is fine rather than cloddy. Second, the soil below the seed must be uncompacted if root growth is to be rapid, and it should contain a readily available supply of nutrients. Finally, the soil above the seed should be loose enough for the delicate shoot to emerge unhindered and begin photosynthesizing (Fig. 5.3).

If the soil has been dug in the autumn and a good frost-mould exists, then a seedbed can be readily prepared as the soil dries in the spring, simply by a light raking. Heavy soils which have been dug in the spring or light soils which have compacted over winter will require more work on them. Loosen the soil to about 4 in. (10 cm) either by forking or better still, because there is less risk of bringing up wet soil,

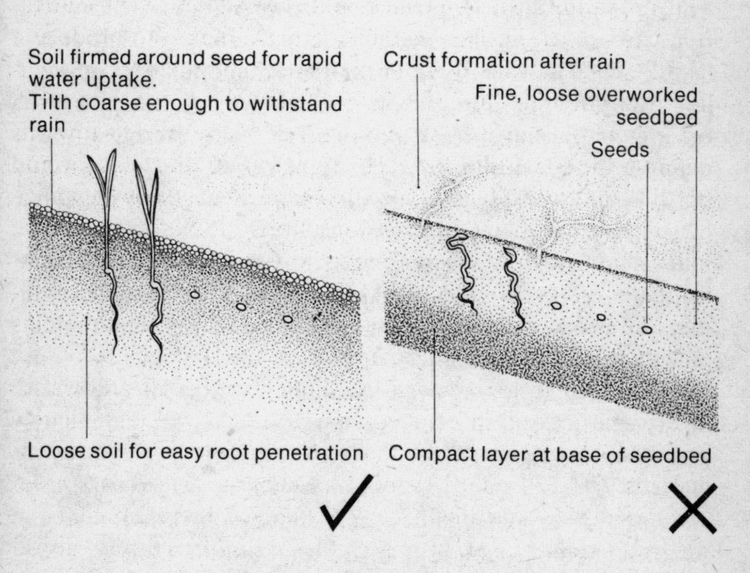

Fig. 5.3. Seedbed preparation.

by using a tined-cultivator. Seedbed preparations should only be attempted when the soil is dry enough for clods to crumble easily. This may mean a few days delay to allow the soil to dry, but you will be rewarded by better crop growth than if the soil was worked wet and became compacted below the depth of sowing. Drying can always be hastened by covering the soil with a cloche for a few days before the seedbed is required. If the soil is too dry to produce a good tilth, watering the area a day or two before final preparation will help the clods to break down.

If the soil has been dug in the spring it may be necessary, particularly on light sands and peats, to firm the soil, preferably by rolling rather than treading. This reduces the risk of surface drying during germination and ensures that the roots come into contact with the water held in the soil pores. In most circumstances treading or rolling is unnecessary and should never be attempted on wet soil.

It is common practice for gardeners to apply and work in fertilizers in the final stages of seedbed preparation. This procedure requires care however since too high a concentration of nitrogen or potassium fertilizer can inhibit seedling emergence by scorching young roots, especially if the soil should dry out. This problem can be overcome by watering or applying only part (up to one-third) of the fertilizer prior to sowing with the remainder applied as a top-dressing between the rows after the crop has emerged.

The final seedbed preparation is carried out by raking, experience suggesting that the aim is to produce a tilth with about 70 per cent of the aggregates ranging in size from that of a grain of rice to that of a pea. In general, the finer the tilth the better but it would be undesirable to work the soil too much since it would be very susceptible to surface capping if heavy rain occured.

If the soil is dry at the time of sowing it is preferable to water to ensure sufficient wetting of the soil rather than to rely on rainfall. Light rain can easily wet the soil to seed depth and initiate germination yet be insufficient to ensure establishment, leaving the partially wetted seeds at risk to fungal attack. If you decide to water it is best to wet the bottom of the drill line thoroughly before sowing and then to cover the seed with dry soil to act as a mulch. This will ensure fast, even germination and emergence under the most adverse conditions.

Providing the soil surface is not excessively wet, carefully firm the drill line with the front of a rake after covering in order to push the seeds into contact with the surrounding soil, especially if the tilth is at all coarse. If subsequent rain causes cap formation, water frequently, with a rose on the watering can, to keep the surface soft. It is a good idea, if the soil is prone to capping, to cover the seeds with a material such as potting compost, vermiculite or perlite which is stable enough not to cap even after heavy rainfall.

Inter-row cultivation

Hoeing is a quick and effective way of killing annual weeds which otherwise would compete with the growing crop for water, nutrients and light. Contrary to popular belief, frequent hoeing to maintain a loose surface tilth or 'dust mulch' has a negligible effect on water loss from the soil. This is primarily because the greatest water loss occurs when the soil is too wet for hoeing. In any case, well-aggregated soils dry rapidly at the surface and are in effect self-mulching. Excessive hoeing can be harmful, by damaging crop roots, bringing wet soil to the surface, and producing a loose surface unstable to rainfall. It is therefore preferable to reduce hoeing to a minimum. Hoe only if weeds are present and keep as shallow as possible.

Many gardeners regard potatoes as a 'cleaning crop', the conventional widely-spaced rows allowing control of perennial weeds by hoeing and by the subsequent cultivations involved in earthing-up. Current commercial practice, with the introduction of effective herbicides, is to control weeds by spraying and to reduce the number of inter-row cultivations to the minimum. Perhaps it is no coincidence that potato yields have increased markedly since this change in practice; gardeners could profitably adopt the same procedure. Potatoes can be successfully grown 'on-the-flat', without earthing-up, and this method is useful for no-diggers where the minimum of soil disturbance is desirable to avoid destruction of the natural structure which has developed. Try planting small tubers weighing about 2 oz. (56 g), 4 in. (10 cm) deep at a spacing of 16 × 17 in. (41 × 43 cm).

A SOIL MANAGEMENT SYSTEM FOR THE GARDEN

In addition to nutritional fertility, soil management is, essentially, concerned with the manipulation of soil structure. The aim of good management is to produce and maintain a stable structure to a good depth, which is free from any defects that would inhibit seed germination and seedling emergence or prevent the free exploitation of the soil by roots. The management problems associated with sands, silts, clays and loams are quite different from each other so that strict adherence to any one system for all soils is fraught with difficulties. Nevertheless, an extension of the bed system, briefly discussed already in connection with drainage, can be recommended and adapted to suit personal circumstances and inclinations.

The main purpose of a bed system is to avoid compaction of the soil by unnecessary trampling. If this system is combined with deep loosening, we have the basis for a system of wide applicability referred to by some gardeners as the 'deep-bed' method (Fig. 5.4). The method begins with double digging to remove any subsoil compaction and to create a structure favourable to free drainage, aeration and root penetration. If available, organic manures should be incorporated at this stage to help preserve the loose structure, improve water retention, encourage soil fauna and increase fertility. Do this operation throughly and it will be unnecessary to repeat it for several years, especially if the structure initially created by cultivation is stabilized by subsequent root residues and biological activity.

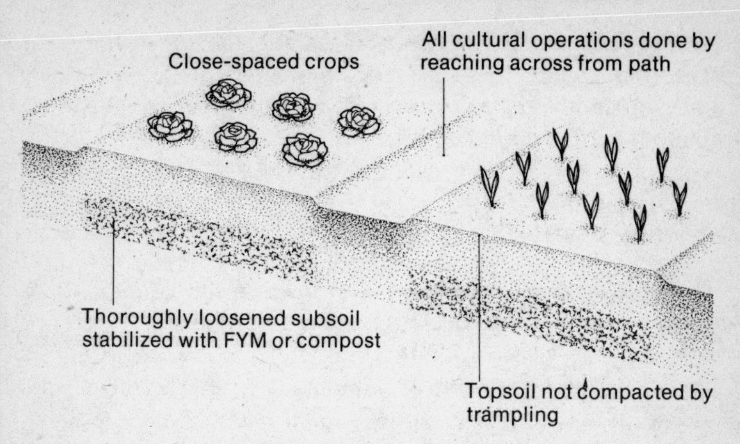

Fig. 5.4. Deep-bed system.

Keep the pathways as narrow as possible (about 18 in. or 46 cm) to avoid wasting ground but wide enough to use a wheelbarrow if required. It is not essential to form a raised bed. Although this has advantages in high rainfall areas and makes reaching into the centre of the bed easier, in dry areas it may lead to an exaggerated loss of water by evaporation from the exposed sides. Even in the latter areas, however, it may be advantageous to have one raised bed which could be used for early sowings, by virtue of its better drainage and earlier warming up in the spring. If you are really keen, sloping the bed 5—10° towards the south will help raise soil temperature.

Since, with this system, the crop can be tended from the paths, the need for the widely-spaced rows formerly required for access is unnecessary. Instead, the crop can be grown at a higher density than usual, and with an even pattern of plant arrangement full use can be made of available land (see *Know and Grow Vegetables*). It is important to remember with this system that *all* cultural operations should be executed from the paths. *Never* walk on the cropped area. If difficulty is experienced in stretching across the bed use a wide board to spread your weight evenly.

After harvesting the crop, give careful thought to the next operation. If the topsoil has not slumped and is free of weeds and crop debris, digging is superfluous and reduced cultivation methods can be adopted.

CROP ROTATION

If some crops are grown year after year on the same patch of ground, yields tend to decrease, even if adequate nutrients are supplied. The reasons for this vary from crop to crop but in most cases can be attributed to an increase of a specific pest or disease in the soil (see *Know and Grow Vegetables*). Since some of these are difficult to eradicate once they are established, it is normally suggested that the same crop, or group of closely related crops, should not be grown repeatedly on the same site. This is good gardening practice, but it does require careful planning if one susceptible crop is not to follow another.

Most gardening books recommend a crop rotation plan which, in addition to disease considerations, is based on nutritional and liming requirements. An example of one rotation frequently suggested is shown below:

Typical 3-year rotation

Add organic manure	*Add fertilizer and lime*	*Add fertilizer*
Group A. Other crops	Group B. Brassicas	Group C. Root crops
Peas	Cabbage	Carrot
Beans	Cauliflower	Parsnip
Onions	Brussels sprout	Beetroot
Leeks	Broccoli	Potatoes
Lettuce	Swede	Tomatoes
Celery	Turnip	

Group sequence

1st year ABC	2nd year BCA	3rd year CAB

The adoption of this type of plan is an effective way of reducing pest and disease problems. However, good soil management may require a more flexible approach to manuring than is commonly suggested. Applying organic manures one year in three is unlikely to increase soil organic matter appreciably. Yet more frequent applications are not suggested because it is widely believed that organic manures will cause root crops such as carrot and parsnip to grow forked or mis-shapen. This is not borne out by experiment. Additions of 10 lb. per square yard (5.4 kg per square metre) of well-rotted FYM had no effect on the root shape of carrots. In fact, many root crops are very responsive to organic manures and it might well be used to advantage at their stage in the rotation.

Apart from this minor criticism of most plans, crop rotation can be regarded as a valuable aid to soil management. It encourages an orderly sequence of cropping so that as one crop or group of crops is harvested, the land they occupied can be immediately prepared for the next. In this way, the need for digging is often reduced. For example, overwintered brassicas can follow early peas with little soil preparation beyond hoeing off any weeds and raking in fertilizer or lime if required. Similarly, ground from which early potatoes have been lifted can easily be levelled for autumn-sown onions or overwintered broad beans.

CONCLUSIONS

Good soil conditions are the basis for the production of high yields of good quality vegetables. Fortunately, even the most unproductive soils can be made to flourish, given time and skilful management. Conversely, bad management can ruin a potentially good soil.

As a guide to good soil management, always observe the following points.

- Ensure that the soil is free-draining. Good drainage is vital to good structure.
- Check acidity and lime if necessary (see *Know and Grow Vegetables*).

- Eliminate compaction by thorough cultivation. Ideally double dig the vegetable plot at least once, even if you are working towards a no-dig system.
- Maintain or improve organic matter levels by using bulky organic manures whenever available. However, do not expect vast improvements unless large quantities are applied regularly.
- Return crop remains to the soil via the compost heap.
- Do not cultivate or walk on soil which is very wet.
- Always do the minimum cultivation necessary to achieve the desired effect. Do not overwork a seedbed as it will exacerbate capping problems.
- Do not hoe unless weeds are present and then keep as shallow as possible.
- Rotate your crops to reduce the chance of a build-up of soil-borne pests and diseases.

6 How vegetables grow and develop

We expect vegetable plants to change as they grow and so provide the part we eat in its familiar form. The pea seedling grows but as it does so it changes and develops flower buds which grow into flowers which, in turn, develop into pods which then grow and are harvested. The stimuli causing the changes in the patterns of growth, which is what development is, can be internal or external. Internal stimuli arise within the plant itself as it grows, whereas external stimuli are changes in the environment in which the crop grows which, in turn, affect the workings of the plant. Important external changes which we will see govern development can be in such factors as the temperature or the length of the light period encountered each day. Knowing how these external and internal stimuli can affect the development of different vegetable crops enables the gardener to use their effects to maximum advantage. If we grow cauliflowers at a spacing of 6 in. × 6 in. (15 cm. × 15 cm) we restrict the *growth* of the individual plants but development proceeds more normally, and small curds, mini-cauliflowers (see p. 66), are produced, each being a one-person portion and suitable for freezing whole. In producing mini-cauliflower, we have altered the normal relationship between growth and development to get what we wanted.

Normally we are most concerned with not allowing growth and development to get out of step. Bolting is one of the consequences of things being out of step. But as we will see with the onion and the cauliflower, we can sometimes obtain useful control by manipulating the relationship between growth and development. First of all we will discuss growth, the basic process which enables us to harvest the sun's energy in a form we can use to provide us with the food energy we need.

GROWTH

All green plants use the process of photosynthesis to fix the sun's energy, making it into the substance and structure of the plant. In the early stages of the growth of seedlings, the rate at which this takes place and the way in which it occurs is similar to compound interest accumulating in a savings bank. The rate of interest is termed the 'relative growth rate' and, of course, the actual amount of interest depends upon the amount invested and the rate of interest.

In the savings bank, interest is usually quoted as some percentage per annum. If it is 10 per cent we know that £100 will earn £10 in a year and that if this is reinvested, which is what plants do, it will earn 10 per cent of £110 = £11 in the second year. Crop physiologists who study growth express interest rate slightly differently and would say that 10 per cent interest was £0.1 per pound per annum. Interest in banks can be paid monthly or even daily on the amount deposited, but in plant growth it is paid instantly so that the interest is immediately put to work to earn more growth. Further, the period of investment is often short, as when a crop of, say, radishes is mature in only a few weeks. So, relative growth rates are quoted as increases in weight, per unit weight, per day. A common rate is 0.1 gramme (g) per day or a 10 per cent interest rate in only a day.

Temperature is one of the main factors limiting the growth of vegetables in the open. The relative growth rate of plants roughly doubles for every 10°C increase in the temperature within the range normally encountered. If you are a pessimist it is equally true to say that it is halved for every 10°C drop in temperature. The effects of this are striking.

If we take the relative growth rate as being 0.1 g. per g. per day, a plant weighing 10 g (about ⅓ oz.) will weigh 11 g after one day, 12.1 g after 2 days, and 19.5 g after a week. If the relative growth rate is doubled by increasing the temperature by 10°C the original 10 g becomes 12 g after one day, 14.4 g after 2 days and 35.8 g after a week. So although we have only doubled the relative growth rate the warmer plants would be 84 per cent heavier than the cooler plants after only a week, and would have increased in weight by 25.8 g as compared with 9.5 g. The value of cloches or frames to increase the temperature is obvious, but another feature of compound interest is that you earn more by having more in the bank.

For example, a 30 g seedling at low temperature will, at the relative growth rate we are assuming, increase its weight by three times that of a 10 g seedling, so in a week the increase will be 28.5 g. This is more growth than our 10 g seedling made in a week at a 10°C higher temperature. It illustrates the value of using transplants, which may only have the same relative growth rate as seeds sown directly, but because they will be bigger at the time of transplanting will earn more interest. Indeed, the bigger seeds from within a packet will similarly show advantages over the smaller seeds just because the bigger seeds have 'more in the bank' to start with. However, these simple rules of growth only apply to seedlings well spaced from each other. In older plants or in younger ones crowded together, other factors soon begin to limit and reduce the relative growth rate.

The reduction in relative growth rate as plants grow is to some extent predetermined just as we stop growing as we get older. However, part of it comes about as a result of more of the daily energy captured from the sun being used to maintain the older lower leaves which become shaded by the younger ones as the plant grows. Similarly, leaves of neighbouring plants can cast shade, and this competition for light reduces relative growth rate. As we shall see, changes in development can also reduce the growth rate by diverting the products of photosynthesis into parts such as bulbs and roots which are not earning their keep by photosynthesizing.

Widely-spaced leafy crops which are harvested at a relatively early stage of what could be their full life-cycle are those which maintain the highest relative growth rates for longest. Lettuce is a good example of such a crop and as such it is amongst the most responsive to temperature. However, we have been considering here the unbridled growth of plants. In the garden all kinds of factors operate to limit growth rates. Two of them we have already identified, namely, low temperature and shading. Vegetable gardening is largely about removing as many of these factors limiting growth as we can. Thus we explained in our first book how to control the pattern of plant arrangement to reduce the competition between plants and how to use fertilizers and water to get maximum growth rates. In this book, Chapter 4 makes it clear how important weed control is in allowing us to achieve the maximum growth of crop plants. This growth is channelled by development into different organs at different times, and by understanding the factors which control development we can manipulate vegetables to get more edible produce from our gardens.

DEVELOPMENT

The mature plant ready for harvest is not simply a bigger version of the seedling. If it is a pea or bean it has developed flowers and fruit during its growth. If it is a lettuce or a cabbage it has developed cupped leaves giving us the familiar hearting – or it may not have done, in that it could have bolted to flower without hearting. What causes this skipping of a stage in the normal development? Can we do anything about it and, if so, what can we do? We can answer these questions for most vegetable crops but the amount of detail available varies. As a prime example of a crop where gardeners can control growth mechanisms, we look first at onions.

Onions

The bulb we eat is the storage organ which gets this biennial plant through to its second year of growth in which it normally flowers and seeds. But let us start with the normal crop grown from seed sown in the spring, using varieties which are now usually of the Rijnsburger type. The emergence of the seedling though the soil and its growth are initially slow because the temperatures are low. As the days get warmer the growth rate of the seedlings increases and more leaves are formed. However, as the days get warmer, so also does the length of the light period each day; this not only acts to increase the growth we get but it also induces developmental changes.

It must be explained that it is not the length of the day that is important but rather the length of the night. In nature, the two go hand-in-hand, long days giving us short nights, but scientists have shown that they can, for example, obtain a long-day effect by interrupting a long night by a short period of low intensity light from ordinary electric light bulbs. In spite of this, scientists continue to talk of long-day and short-day effects, and here we are concerned with the effect of long days on our onion seedlings.

With the spring-sown bulb onions grown in Britain, a day-length of 16 hours stops the production of ordinary leaves and instead the fleshy scale-leaves which make up the bulb are formed. Once the plant has been programmed in this way, it is irreversibly committed to bulbing and to ripening. If only a small amount of green leaf per plant has been produced before bulbing begins then the onion bulb is bound to be small. A small amount of foliage can only produce a small bulb. It follows that getting the maximum amount of foliage before the long days cause the switch to bulbing is of great importance if good yields are to be obtained. One way to do this would be to increase the temperature early in the growth by protecting the seedlings under cloches, or transplants could be raised in mild heat to give bigger plants earlier. Most commercial onion growers rely on early sowing but amateurs often favour using onion sets.

Onion sets are very small onion bulbs which are stored and planted the following year and it follows from what has been said that they are produced by late sowing. In Britain, seeds are sown in about the third week in May, only a month before the day-length reaches its maximum. Bulbing therefore takes place with very little leaf on each plant, so small bulbs are formed. Because we know they are only going to be small they are sown very thickly, much as one would sow seed in a seed-box to raise young plants. Indeed, the amateur can often raise enough sets for his own use in one or two seed-boxes; this has the advantage that they can be put in a frame or glasshouse to assist their ripening.

The ripened set, just like the larger bulb grown in the garden, is at first dormant. Unfortunately, dormancy is readily broken, otherwise it would be easy to store both sets and bulbs. The factor which most commonly breaks dormancy early in the bulb's life is water. Ripe bulbs or ripe sets left out in the rain will soon begin to sprout. However, if they are kept dry, temperature is the major factor controlling dormancy of bulbs. Dormancy is preserved at both high (about 70°F, 21°C) and low (about 32°F, 0°C) temperatures and is broken at intermediate temperatures, particularly those around 50°F (10°C). Once dormancy is broken, any reasonable temperature will induce sprouting and sprouted bulbs soon deteriorate.

Because the onion is a biennial plant, if we planted large bulbs that we had stored, we would get flowering plants the following year. This is how the seedsman raises seed. But with sets we do not want flowers, for we have grown the set to give us a start in the race to produce plenty of leaf before the days lengthen and induce bulbing. Sets do often give at least a proportion of flowering plants and their yield is lost. How can this be avoided?

The low temperatures of winter start a developmental response in the stored bulbs causing them to form flower buds inside the bulb. Fortunately, small bulbs are too 'young' to indulge in this change to flowering. They are often spoken of as being in the juvenile phase and not having reached puberty. They are not large enough for sexual reproduction even although they receive a stimulus that would provoke bigger bulbs to form flowers. Varieties used for sets generally have a slightly bigger bulb size at puberty than ordinary varieties but it is unwise to use sets larger than 15 mm (0.6 in.) in diameter unless they have been 'heat-treated'.

Heat-treatment prevents flower formation in bulbs which would otherwise be susceptible, by denying them the cold stimulus they would need. The sets must be stored throughout the winter at 87° to 98°F (25° to 30°C), temperatures usually only found in the home in airing cupboards. Because of exposure to low temperatures in transit and in the shops, it is usually safest, and more economical, to buy small sets. The alternative way of getting a good start is to sow seeds in the autumn.

Autumn-sown onions fall into two groups. With traditional long-day varieties, autumn sowing in protected conditions is used to produce transplants for putting out in the spring. The bulbs produced in this way are larger but very little earlier than those from spring-sown seeds because the plants have to wait for the long-day (16 hour) stimulus before bulbing can start. With the newer, intermediate-day-length varieties, which we introduced largely from Japan, the seed can be sown in the open where the crop is to be grown and no transplanting needs to be done. Very early crops in June can be obtained from some of these varieties, because they can *start* to bulb when the days are 12 hours long – at the end of March. From a developmental point of view the problem with both groups is the same – to avoid bolting.

We have already seen that bulbs bigger than 15 mm (0.6 in.) in diameter will form flower buds in response to low temperatures. Onion plants also behave similarly in that small plants will not respond to the stimulus of the low winter temperatures and so do not flower in the following year. In raising plants under protection, either by autumn sowing or very early sowing (January in Britain), they can be protected from the low temperatures that would trigger off flowering. But in the open we have to rely on not getting the plants too big by avoiding sowing too early. The problem is that if sowings are made too late the plants may be too small to survive the rigours of winter.

For this reason very precise sowing dates have been determined by experiment for autumn sowing the Japanese type of onion in different parts of Britain. In the south of England the last week in August was found to be reliable, whereas in the Midlands sowing should be made in mid-August, and in the north and Scotland in the first week of August. It follows that you should not just sow but you should also water, if necessary, to get rapid germination and the plants started at the right time.

Because autumn sowing gives us a long if somewhat cold period for leaf growth, we can get good yields from varieties that bulb in days much shorter than the traditional varieties grown in Britain. The earliest varieties will start to bulb when the day is only 12 hours long and so will give ripe bulbs in early June. Varieties bulbing in 13 to 14 hours are also available giving later ripening but also higher yields. Unfortunately, most of these Japanese types do not keep well so they should be grown for summer use within a month or two of harvest. One peculiarity of these autumn-sown onions should be noted here and that is that they bolt less if they are kept well supplied with nitrogen fertilizers during the winter months. At present we do not understand why this should be so but experiments have shown it to be most important in reducing bolting.

If you enter for shows you may want to save your own seed from a prized bulb. Generally, this is not a wise move as saving seed from one or only a few plants could weaken the strain, but you could try growing from 'pips'.

Pips are small bulblets which sometimes form naturally between the flower stalks of the globular flower heads of onions and leeks. Pot leek growers, those who grow mammoth leeks for show, particularly in the north-east of England, nearly always grow their prize specimens from such pips treating them like onion sets. You can induce the formation of pips in the flower head of onions and leeks by shaving off the flower buds with a razor blade just before they open. Be sure to leave the flower stalks, and in due course you will find the little bulblets forming. They ripen like the true bulbs on an onion.

Having looked at onions in detail, we can now use them as a

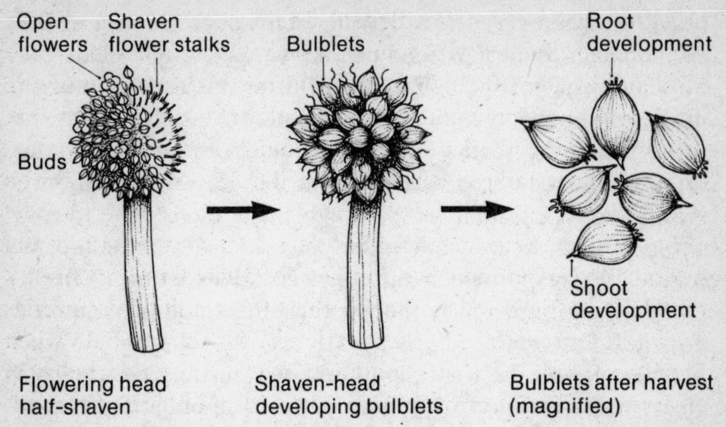

Fig. 6.1. The flower heads of onions and leeks can be made to produce bulblets ('pips') by cutting off the buds just before they open. The bulblets are about $\frac{1}{4}$ in. (6 mm) long and can be planted like sets.

basis for understanding the development of other vegetables as it affects us as vegetable gardeners.

What causes plants to bolt?

'Running to flower' when there should not be any flowers is referred to by gardeners as 'bolting' and sometimes the plants will be said to have 'shot'. Because bolting stops root swelling and tends to make what is there rather woody, it can be very annoying to find all the beetroot going up to flower just when it looked as though the roots were reaching a harvestable size.

Many vegetables, like the onion described above, are biennials. This strictly means that they require a cold stimulus to start the formation of flower buds but they are not sensitive in this way to cold when they are seedlings or young plants. The onion we have already described is therefore a classical biennial. There are also vegetables which behave similarly but are classed rather as winter annuals in that even the seed that has taken up water but has not yet started to grow is receptive to low temperatures as a stimulus for flower initiation.

Nature is never quite as tidy as scientists would like it to be so we find that there are examples of intermediate groups between true winter annuals and true biennials. Turnips are winter annuals in that their seed can be 'vernalized' by being put in the cold in an imbibed state. (This is not to be confused with 'stratification' which involves exactly the same treatment but is done to overcome dormancy in seeds, often of trees and shrubs, and does not induce flowering.) Celery appears to be a transitional type in that very young plants are receptive to cold but are not as receptive as older plants. However, before we consider the responses of different vegetables let us see if we can define more precisely the temperature conditions required to initiate flowering.

Generally, the best temperatures for *vernalization*, the cold effect on flowering, are within a degree or so of 39°F (4°C). At lower temperatures processes get slowed down too much, and at temperatures above about 54° to 58°F (12° to 14°C) there is usually no stimulus to flower. Indeed, provided higher temperatures occur soon after vernalizing temperatures, the vernalization effect of the lower temperatures is often cancelled out. Thus cold nights followed by warm days tend not to count in clocking-up the stimulus to flower in biennials or winter annuals. But if the days are also cool, the plants are set along the path to flowering. As a very rough guide it takes only about six weeks at low temperatures to induce the initiation of flowers, but you may not see the effect until several weeks later. Usually we *see* bolting in early summer but the cause was invariably the cold in early spring. We have already dealt with onions so now let us consider other crops more specifically.

Bolting in root crops

Turnips. We have already noted that the seed of this crop is sensitive to cold so they rate as one of the most vulnerable crops; the only answer is to delay sowing or raise early crops under the protection of cloches or frames where the temperature will be higher than in the open.

Beetroot is the next most sensitive root crop where the usual recommendation is to delay sowing into late spring. The imbibed seeds are not sensitive to cold but the young plants are sensitive. Incidentally, we were unable to show that transplanting red beet encouraged bolting although a lot of gardeners seem to consider that it does. Could the explanation be that they tend to transplant the biggest seedlings which are more sensitive to low temperature than smaller plants? You will probably have noticed that only some of the plants from an early sowing bolt. The ones that do not could be younger in that they have come up later and hence have clocked up less cold stimulus. However, it is equally likely that the non-bolters were inherently more resistant to bolting. In beetroot these genetic differences in the tendency to bolt have been exploited by breeders and at least two varieties are on the market, Avonearly and Boltardy, which can be sown in early spring with much reduced chances of bolting.

Radish readily run to seed in the long days of summer. This is because it is an annual that only requires long days to make it flower. Seedsmen have selected types suitable for sowing at the different seasons of the year, and it is important to use varieties appropriate to their season. The summer varieties will form edible radishes before 'bolting' provided they are well nourished and watered. In poor conditions they will bolt without forming proper radishes first.

Carrots of the modern varieties rarely bolt as a result of early sowing but they will frequently do so if autumn sowing is attempted. As far as is known, the young plant with one true leaf is the minimum size that can be vernalized.

Parsnips have not been studied in detail but they seem even less likely to bolt than carrots. The same cannot be said for celery and celeriac which belong to the same family.

Bolting in leaf crops

Celery and celeriac are very sensitive to cold and become much more so as the seedlings increase in size. Practical experience suggests that maximum sensitivity occurs at about the plant size usually aimed at for transplants. There is a grave danger with these crops that keenness to slow down rather advanced seedlings by hardening them off can lead to a high proportion of bolting plants a little later. If you really must slow down some transplants try clipping them with scissors to about 3 in. (8 cm) high and do not put them outside until it is warm. The clipping will give you sturdy plants which, in experiments with celery, have transplanted better.

Cabbages and Brussels sprouts are quite big seedlings before they respond to low temperatures. Like onions, they have a clear-cut 'puberty' in that no amount of cold applied to young plants will cause them to bolt. Generally, puberty coincides with the top of the stems becoming about as thick as a pencil.

It is very unusual for bolting to be a problem in Brussels sprouts, summer cabbage, or winter cabbage, although the latter will obviously run to seed in the spring. The major problems are usually encountered with spring cabbage, which is a crop which is always bound to be on a razor's edge as far as bolting is concerned.

As with overwintered onions, already discussed, an important step in controlling bolting in spring cabbage is to choose the correct sowing date. About 20 July is right in the Midlands of England with sowing being a week earlier in the north and a week later in the south. Plant breeders have done a great deal to reduce the risks of bolting and our variety, Avon Crest, is particularly resistant to bolting and can be left until later in the spring to form good hearts if required. If you see signs that your spring cabbage crop is going to bolt, then the best thing you can do is to eat it quickly and make a note to sow a little later next year.

It is useful to realize that mild autumns and winters are the worst conditions for bolting in both spring cabbage and autumn-sown onions. The mildness allows growth which advances the plants on their path to puberty and even mild winters are cold enough for good vernalization – indeed they can be better in this respect than really cold weather. So do not be tempted to put cloches too early over spring cabbage, cover them only when the spring weather begins to arrive.

Chinese cabbage is becoming increasingly popular, but the main problem with its culture under British conditions is the tendency to bolt. Germinating seed and plants are triggered by temperatures below about 50°F (10°C) to form flower buds and our long days of early summer enhance this effect of low temperature. Some varieties, notably some of the new varieties from Japan that are now in trials, are less prone to bolting than others but even with such varieties it is unwise to plant transplants raised at temperatures above 45°F (7°C) before mid-May, or to make sowings out-of-doors before mid-June.

Lettuce varieties usually grown in the open will only bolt after hearting. They are responding to the long days of summer in doing this and when the days are longest and warmest there is often only a matter of days between a heart being mature and it beginning to grow out into a flower stalk. The varieties adapted for growing under glass in the winter are different in that they will bolt without hearting if grown in long days. So do not be tempted to use one of these winter glasshouse types for your summer crop.

Spinach is known to scientists as one of the most sensitive plants to long days. Only one long day of 13 hours duration will induce some varieties to flower, the male and female flowers being on separate plants. Thus, the best seasons for the production of spinach are in the shorter days of the autumn, winter and spring, although some protection is usually needed for crops during the coldest periods. These crops in short days will produce leaves over a long period and can be relied on to give several harvests. For summer use, a succession of sowings has to be made as each sowing will usually only give one picking before bolting stops leaf production.

Spinach beet sometimes also called 'Silver Beet' or 'Perpetual Spinach', is a form of beetroot which has been bred for its spinach-like leaves. It only 'bolts' in similar conditions to beetroot and hence can provide a good source of summer spinach without fear of bolting, provided it is not sown so early that it receives the cold stimulus it needs for flowering.

FLOWERING

Up to now we have been dealing with vegetables that most gardeners never want to see in flower. We have seen how to control it and so reduce its devastating effects on yield. Now we turn to those vegetables that we do wish to see flower or, in the case of the first group, to form flower buds.

Cauliflower, calabrese and broccoli. The cauliflower curd, the part we eat, is really a mass of partially-developed flower buds. This is more obvious in sprouting broccoli and calabrese. Low temperature is still the stimulus for bud formation in all this group and, like cabbages and onions, puberty has to be reached before the plants are sensitive, in this developmental way, to cold. The temperature responses of calabrese and broccoli are similar to cauliflower, which shows an enormous diversity between types in the temperatures to which they respond.

The winter cauliflower types need the lowest temperatures for the longest time to start flower formation and this prevents them producing curds in the autumn. On the other hand, some varieties of summer cauliflower can form curds when the temperatures are as high as 70°F (21°C). So as a group, cauliflower and its relatives do not obey the temperature rules given for root crops, cabbage and Brussels sprouts, and onions.

Sometimes quite young cauliflower plants will form curds which are very small (about 2 in. (5 cm) in diameter). These are edible but they are, of course, disappointing as compared with the full-sized curds expected. It is usually only the earliest cauliflower varieties that suffer from this fault which is known as 'buttoning'. These varieties produce relatively few leaves before they form a curd and if for any reason these leaves are smaller than they should be, the plants cannot produce large curds, only 'buttons'. The main factors causing the leaves to be smaller than normal can be summarized as ones which produce a check to normal growth. Any such check seems to predispose plants of these early varieties to buttoning. A period of low temperatures, a shortage of water, exposure to wind and a shortage of nutrients are all factors which have been associated with buttoning.

With the summer and later groups of cauliflowers the best quality and head size have been obtained by transplanting no later than five to six weeks after sowing. These relatively young transplants give crops that are more uniform in their maturity and higher yielding than crops from older transplants.

Experiments at Wellesbourne have shown that cold treatment of cauliflower transplants can also improve the uniformity of maturity of crops. The treatment 'saturates' the cold requirement of all the plants making sure that any natural variability in their cold requirement is swamped. It can be done in a domestic refrigerator by putting the transplants into a polythene bag and placing them in the refrigerator in a position where they will not freeze. Usually, the compartment recommended for the storage of salad vegetables is satisfactory. They will not take harm for up to 21 days but usually 14 days is sufficient to ensure the desired effect is obtained.

It is unusual for the gardener to want all his plants from one sowing to mature within a short period, although if the intention is to deep-freeze florets for later use this can be convenient. For fresh consumption a good spread of harvesting from one sowing is what is needed. This can be achieved by transplanting some plants from the seedbed when they are five to six weeks old and storing others in the refrigerator, as described, for planting at weekly intervals over the next three weeks.

Cold-stored plants should not be transplanted when it is hot and sunny. It is best to try to choose a cool, dull day, but if this is impossible shade the transplants and sprinkle them with water for the first few days.

Peas and beans all flower without any environmental stimulus. They are inherently programmed to flower and are mostly true annuals in that, having flowered and produced seed, they then die. The exception is the runner bean which is a true perennial, although it is usually treated as an annual because its fleshy roots, which can survive from year to year, are usually killed by frosts. If you lift them in the early autumn and store them like dahlia tubers in moist sand or peat in a frost-free place, then you can plant them in late spring and stand a good chance of an earlier crop.

If you do decide to try this, be careful to save only the roots of healthy plants. You run the risk of accumulating diseases over the years if you continue to use this method of propagation.

To return to flowering, the most usual complaint of gardeners about peas and beans is not that they are short of flowers, but rather that many of the flowers fail to produce pods. In general, this failure of many flowers to set is entirely natural as once the plant has set some seed it concentrates its resources on those pods to ensure the survival of the species. The extent of the pod-load that a plant will stand depends very much on the plant's size. For example, when commercial varieties of peas are grown at close spacings, say, 6 per square foot (60 per square metre) the individual plants are small and each will only bear 3–5 pods that contribute to yield. Indeed, flowering stops once this number of pods has been set. At wider spacings, say, 2 per square foot (20 per square metre), the individual plants are larger, so that one will and can bear more pods. This is unlikely to be enough to keep the yield per unit area comparable to that obtained at the closer spacing, but it would be more per plant, probably 6–10, and you would notice that the flowering period went on for longer. The same holds true for French beans, but with broad beans and runner beans there are complications.

The peas and French bean flowers pollinate themselves, whereas the broad beans and runner beans need insects to pollinate them. Thus, if there are no insects about to do the job, flowers of broad and runner beans will drop off even though the pod-load is not restricting them. Fortunately, they keep producing more flowers under these circumstances and sooner or later pod-set is obtained.

Experiments have clearly shown that syringing runner bean flowers with water does nothing to promote their setting pods. Warm weather that promotes plenty of insect activity is the only thing that will ensure the set of those first runner bean flowers. Thereafter, the most important factor which will help to give continuity of cropping is regular, frequent and thorough picking. If you see any pods that you missed when picking earlier and that have, as a result, got so big and tough as to be inedible, pick them off and throw them away. If you leave them on the plant it will only cause the flower buds or very young pods to drop off. Of course, poor growing conditions and, in particular, conditions that are too dry, will also cause the plant to carry a much reduced pod-load. So it is important that the roots are kept moist and proper attention has been paid to the nutritional needs of the crop. If this is done and you pick thoroughly and frequently, you should get good crops and a long picking period.

Recently, seedsmen have re-introduced white- and pink-flowered varieties of runner beans which are claimed to set pods more readily than the scarlet-flowered types. We have no evidence that would support this view.

Tomatoes never fail to produce flowers but sometimes the flowers on the first truss fail to open because the plant is growing so fast that they are bypassed and abort. More commonly, the flowers, particularly on the earliest trusses of outdoor crops, may fail to set. This is most likely to be because it has been too cold for the pollen to grow and fertilize the flower. The best temperature for the pollination of tomatoes is about 77°F (25°C), while at temperatures of about 41°F (5°C) pollination ceases. It is important to appreciate that we are talking here of the growth of the pollen needed to fertilize the flower. The transfer of pollen can sometimes cause problems in the still atmosphere of a glasshouse and shaking the plants to assist transfer is sensible. However, in the open, pollen transfer is not limiting but low temperatures, particularly at night, can and do limit early fruit-set. Some newer, outdoor, bush varieties of tomato have been bred to have pollen which will be effective at temperatures as low as 6°C (43°F).

It is possible, however, to substitute for the pollen by spraying the open flowers with hormones. These stimulate the growth of the fruit even at temperatures too low for pollen to work, but the fruits will be seedless. It is important to follow the directions on dilution and application with great accuracy as hormones are likely to produce odd results if care is not taken. Properly used, they are entirely reliable and can be very helpful in giving you earlier tomatoes.

Marrows, courgettes, cucumbers, ridge-cucumbers and gherkins all flower readily but have separate male and female flowers. Naturally, only the female flowers bear fruit. In all this group there is a tendency for the proportion of female flowers to increase as the days get longer. Gardeners often get frustrated by finding that the first flowers on their marrows are all male but this is only a manifestation of the lingering effect of the shorter days of early summer and very quickly female flowers are formed.

Marrows are the only ones in this group that may need some help with pollination. Pick off the male flowers when they are fully open, strip off the petals and thrust the pointed mass of anthers into the centre of the female flower. You can leave it there or pollinate more than one female flower with it by repeating the operation. Only pollination will ensure the growth of the marrow. Female flowers are easily recognized in all this group by the minifruit immediately behind the petals. The males have no such small fruits.

The only other problem is to ensure that frame-type or glasshouse-type cucumbers do not get pollinated. With most of the modern varieties this is no problem as they have been bred to produce only female flowers. If you choose to use one of the old varieties that produces male flowers you should remove them before they open so that insects do not get a chance to pollinate for you. It seems strange to be preventing pollination, but pollinated cucumber flowers tend to give flask-shaped fruits which are often bitter to eat and, of course, have seeds in them. The fruit has been bred to develop without pollination into the cucumber we know and enjoy.

It is sometimes recommended to pinch out the growing tip of young cucumber plants to induce branching. The plant then spreads out better to fill a square frame. This leads us now to consider what we are doing when we nip out growing points.

APICAL DOMINANCE AND CONTROLLING GROWTH

Although it is easy to nip off the growing tip of a shoot the consequences of doing so are complex and have been much studied by scientists.

Apical dominance is the scientific term for the control that the primary growing tip of a shoot exerts over the growth of the lateral buds. Those buds nearest the growing tip are dominated by it and do not grow. Such buds are, of course, the younger ones so it does not seem too unexpected that they do not immediately produce shoots. It also seems reasonable to suppose that the growing tip of the shoot takes priority for the supply of materials for growth.

As usual, there is some truth in these commonsense deductions but they are far from the whole story. Scientists have long ago shown that small amounts of a chemical substance called auxin is produced at the tip of growing shoots and in passing down the shoot it stops the leaf buds growing. As it passes down the shoot it becomes diluted and eventually is no longer effective, so the lower leaf buds can grow out and produce shoots.

Thus, if we want to make leaf buds grow and we know that the growing tip of the shoot is preventing them, we can nip out the growing tip and remove its dominance. It follows that we only need to remove the smallest tip which actually contains the growing point. We can leave all the young leaves as these do not exert any influence other than being useful in producing wanted components for growth.

Plants differ in the degree of apical dominance they exhibit. Any plant that produces a single stem with no branches, like a sunflower, has strong apical dominance. We see weak apical dominance in plants that are typically bushy in habit like sprouting broccoli. Onions have been bred for strong apical dominance in that usually one plant gives us only one bulb; occasionally a plant produces two or three bulbs within an outer casing which looks like one bulb. This is because some of the buds, which are present at the base of each scale, have grown and, in turn, produced a bulb. The dominance of the main growing point has been upset in some way, often by damage, to allow this to happen. However, in shallots we see a multiple sprouting from planted bulbs in contrast to the single sprouting we usually get from onion sets. In the shallot the lack of apical dominance in the bulb after storage is contrasted with the normal onion-like apical dominance seen in each daughter shallot as it grows. This serves to show us how finely-tuned these systems have become as a result of hundreds of years of selection.

Brussels sprouts are above all other vegetable crops in exhibiting apical dominance on a grand scale. The leaf buds, the sprouts themselves, are large and important to us as the part we eat. We know the bottom sprouts will be ready to harvest first and now we recognize that one reason for this is that they are furthest away from the main growing point of the plant. If we want the upper sprouts to grow faster we can encourage this by removing the growing tip of the main stem. This is often done commercially to enable all the sprouts on a plant to be harvested at one time, so reducing harvesting costs and, in particular, making it possible to mechanize harvesting. Normally in the garden the cut-and-come-again progression of sprout growth on each plant is a blessing as it gives continuity of supply. But where, say, pigeons are troublesome in the winter there is a lot to be said for harvesting all the sprouts early and keeping the supply going by deep freezing.

If you remove the growing tip when the bottom sprouts are about the size of your little finger nail you will find that in due course you will have all the sprouts on the plant ready for a single harvest. This 'stopping', as it is termed, must be done at the right stage of growth as defined above, and before 1 October. If you leave it after that there is not enough good growing weather to get the effect required.

KEEPING GROWTH IN STEP WITH DEVELOPMENT

Scientists have found it convenient to recognize that the age measured in normal time, or the size of a plant, is not sufficient to describe it fully, so in addition they refer to its 'physiological age'. This is often a term used rather loosely, so it can mean different things depending on its context. Generally, though, plants are regarded as being of the same physiological age when they are at the same stage of development. Puberty (see p. 176) is an age-point on the time-scale of physiological age, as is flowering and seed ripening. There is one popular vegetable crop where the manipulation of physiological age is important to the gardener. This crop is the potato.

Potatoes are grown from seed tubers and it is their physiological age which has marked effects on yield and earliness. In brief, the 'younger' the seed tubers the lower is the early yield and if they are very young even final maincrop yield may be reduced. 'Middle-aged' seed tubers give the best maincrop yields and physiologically 'old' seed tubers the best early yield (Chapter 2 p. 78). But how do we manipulate 'age' when the seed tubers are all bought at the shop and all planted within a month or so of each other?

All seed tubers are dormant when harvested and remain so for periods that vary with variety and the year and site of their growth. Once dormancy is over sprouts begin to appear as very tiny specks around the 'eyes' of the seed tuber. The warmer the conditions after dormancy has ended the quicker the sprouts will begin to grow, and it has been found that the length of the longest sprout, *when sprouting has taken place in daylight*, is a good measure of physiological age. Sprouting in daylight is most important since if kept in the dark the sprouts elongate excessively, are very fragile and their length is less representative of their 'age'. At temperatures below 39°F (4°C) sprout growth stops in most varieties, so such low temperatures can be used to keep seed 'young'.

Any temperature above 39°F (4°C) causes growth of the sprouts and experiments have shown that the length of the longest sprout is directly related to the number of day-degrees above 4°C. This concept of day-degrees is also discussed in Chapter 2 (p. 47) but quite simply a day at 8°C is a day at 4 degrees above the lowest temperature for growth and counts as 4 day-degrees. Thus, a day at 6°C (2 day-degrees), followed by a day at 10°C (6 day-degrees) followed by a day at 4°C (0 day-degrees) would give a total of 8 day-degrees.

With the old variety Home Guard harvested on 15 June, the highest yields were obtained with 'old' seed that had encountered about 1,000 day-degrees since harvest and had sprouts 80 mm long. These yields were about 30 per cent more than those obtained from exactly similar seed tubers which had only encountered 100 day-degrees and had sprouts about 5 mm long. By 30 June when another harvest was taken, the yields from 'old' and 'young' seed tubers were similar but 'middle-aged' seed tubers, which had encountered about 500 day-degrees and had sprouts 40 mm long at planting, were giving yields 27 per cent higher than either the 'younger' or 'older' seed.

Later still, on 18 July, the 'middle-aged' seed was still the best but now the 'younger' seed was only 13 per cent lower yielding. However, the 'oldest' seed was poorer yielding giving 17 per cent less than the 'middle-aged' seed. The plants from the old seed stopped growth and died down first and this is why they became the lowest yielding.

These results have been quoted in some detail as this whole subject is still in its infancy and it is slightly dangerous to assume that the same is true of all modern varieties and other crops in other places. However, the broad effects of advanced physiological age are becoming clear and can be listed as:

earlier emergence after planting
earlier commencement of tuber formation
fewer tubers developing per plant
smaller final plant size and a slower growth of the tubers
an increased susceptibility to drought
more nitrogen fertilizer needed to get maximum yield
earlier die-down of the tops

These points are mostly self-evident summaries of behaviour that was exemplified by the results quoted with Home Guard. The high early yield from 'old' seed tubers is because they start to develop tubers earlier and concentrate their resources on relatively few tubers. However, late yield is poorer from old tubers because they are fewer, individually grow more slowly, and the plant dies sooner. The susceptibility to drought appears to be associated with a less extensive root system.

The preliminary indications with the maincrop variety Desiree are that an 'age' of 1,000 day-degrees is about the optimum for maximum yield at the end of the season.

Seed can start to 'age' before you buy it, so for maximum control of physiological age you should aim to purchase your seed as early as possible. If you can see the sprouts beginning to grow as tiny greenish white dots around the eyes, then you can be sure that the dormancy has finished and that 'age' is beginning to be clocked up. Remember that the guidance given on sprout length for Home Guard is for the length of the longest sprout when it has grown with enough light to prevent it elongating unduly.

Gardening is essentially the practical control of growth and development of cherished plants by manipulating their environment. The more you know about the factors affecting growth and development, the more successful you should be.

Index

OXFORD PAPERBACKS

A Book of Honey

Eva Crane

Dip into this book for facts about bees and honey. Dr Eva Crane, Director of the International Bee Research Association, writes about the history of bee-keeping, and the folklore that has grown up around bees and honey; about honey as a gastronomic delicacy (illustrated with some delicious recipes), as a remedy, and in cosmetics. The book will appeal to apiculturists, botanists, cooks, and anybody who enjoys eating honey.

'An unremittingly fascinating compendium of apian information' *Sunday Times*

OXFORD PAPERBACKS

The English Gardener

William Cobbett

Introduction by Anthony Huxley

'It is a most satisfying read on early gardening, much of Cobbett's advice being as relevant today as it was when he wrote it'. *Yorkshire Evening Post*

'I was surprised and soon delighted at finding *The English Gardener* in paperback . . . (Cobbett's) earthy writing is a reminder that much about the past has still to be rediscovered.' *Country Life*

'A masterpiece of general gardening' *The Observer*

OXFORD PAPERBACKS

Thought for Food

Margaret Sumner

Do processed foods contain any goodness? Does sugar do you any harm? Do you need vitamin pills? If you want common-sense answers to the important questions about the food you eat, read this book. It gives detailed, clearly presented information about the nutritional values of different foods, special diets and dieting, and sensible advice about how your eating habits can be improved.

'A clear, lively run-down on nutrition, with Amazing Facts to silence other dinner guests with' Jocasta Innes, *Sunday Times*

OXFORD PAPERBACKS

Mushrooms and Toadstools: A Field Guide

Geoffrey G. Kibby

Illustrated by Sean Milne

This book will supply the need for a really effective identification guide to British and European fungi. It is more than a guide, however, because the first thirty pages lead from a survey of the whole world of fungi – 50,000 species from microscopic yeasts to giant puffballs – to a consideration of the mushrooms and toadstools which form the basis of the guide. There are sections dealing with the structure, classification, and reproduction of fungi as well as information on their uses and warning paragraphs on those poisonous species to avoid. The identification section consists of 100 pages of full-colour drawings, each page faced by detailed descriptions of the species illustrated, in taxonomic order.

'A competent book in which the lifelike artwork is a real pleasure to look at, and which ably covers a large proportion of the fungi any amateur is likely to find.' *Country Life*

'I find this book attractive with a major strength in its layout and beautifully clear printing of both colour and text' Julia Grollman, *New Scientist*